ASSISTIVE TECHNOLOGIES

Edited by **Fernando A. Auat Cheein**

Assistive Technologies
Edited by Fernando A. Auat Cheein

Published by InTech

Janeza Trdine 9, 51000 Rijeka, Croatia

Publishing Process Manager Jana Sertic

Technical Editor Teodora Smiljanic

Cover Designer InTech Design Team

Additional hard copies can be obtained from orders@intechopen.com

Assistive Technologies, Edited by Fernando A. Auat Cheein
p. cm.
ISBN 978-953-51-0348-6

For EU product safety concerns:
IN TECH d.o.o., Prolaz Marije Krucifikse Kozulić 3, 51000 Rijeka, Croatia,
info@intechopen.com or visit our website at intechopen.com.

Meet the editor

Dr. Fernando A. Auat Cheein is an Assistant Professor at the Federico Santa Maria Technical University, in Valparaíso, Chile, since 2011. He received his degree in Electronics Engineering -with honors- from the National University of Tucumán, Argentina, in 2002, the Master and the Doctorate degree in Control Systems from the National University of San Juan, Argentina, in 2005 and 2009 respectively, obtaining in both, the excellence grade. His Master's thesis concerned the development of Brain-Computer Interfaces (BCI) for controlling robotic devices (robot manipulators, mobile robots and robotic wheelchairs). His doctorate's thesis focused on SLAM (Simultaneous Localization and Mapping) algorithms applied to mobile vehicles, including robotic wheelchairs. He has several publications in both the robotics field and the rehabilitation field. Several of his publications have been awarded as selected or best papers. He is currently a reviewer of numerous indexed journals, conferences, and part of the editorial board of the International Journal of Advanced Robotic Systems (InTech).

Contents

Preface

Assistive technologies aim to enhance the daily life of people with disabilities, improving aspects such as mobility, communication, social interaction, manipulation of several objects and devices, etc. The main importance of assistive technology is based on the idea of making reachable what is not accessible for many people with disabilities. Such is the case of robotic wheelchairs, which improve mobility by using control system techniques, artificial intelligence and advanced sensors; human-computer interfaces (such as brain-computer interfaces) for the generation of commands used for controlling devices, or voice-based systems that reads text or write text on the screen, among other examples.

With this insight, this book offers the most recent advances in the assistive technology field through the contribution of individual chapters authored by eminent professionals and known researchers in their respective topics. *Assistive Technologies* attempts to provide a state of the art and original work made by the authors. This book covers topics from communication to mobility and social interactions, emphasizing the importance of the proposed technology in the improvement of the patient's daily life and offers the reader a comprehensive understanding of the field. In addition, this book is not only aimed at engineers or scientists, but also educators and professionals in the rehabilitation field.

I would like to express my gratitude to the authors for their valuable contribution, and I expect that this book will help to attract more researchers in order to keep developing more and better assistive technology for all.

Dr. Fernando A. Auat Cheein
Department of Electronics Engineering,
Federico Santa Maria Technical University,
Valparaiso,
Chile

Part 1

Education

1

How to Use Low-Cost Devices as Teaching Materials for Children with Different Disabilities

Chien-Yu Lin
National University of Tainan
Taiwan

1. Introduction

This chapter focuses on how to use assistive technology to help children with disabilities. The real situation is that many classes do not have enough resources to buy much equipment for children with different needs. We will introduce how to use low-cost devices to make a simple interactive whiteboard, a low-cost AirMouse and interactive feedback for rehabilitation treatment. Some research has combined the use of a Wii hand controller and infrared (IR) emitter to create a low-cost interactive whiteboard; we applied and extended the relative issues in this chapter that utilizes the Wii hand controller as a key of assistive technology. The contents are applied in Flash software or PowerPoint, so that resource teachers could design the teaching materials without any training.

Computer-aided instruction is widely used in special education and assistive technology. In fact, teachers sometimes may have trouble when teaching their students to use different computer tools (Shimizu & McDonough, 2006), but advances in computer technology have meant replacing translation of traditional paper questionnaires with novel display versions (Hung et al., 2010). Although with the progress of technology the functions of interface have become more complicated day by day, the purpose of this chapter is not only to help children through new teaching materials to enhance their motivation, but also to consider teachers' work load and feelings.

Through the assistance of assistive technology and multimedia design, teachers have enough ability to produce the learning materials of custom-made design in order to support learning by disabled students (Kawate et al., 2009). User-friendly design is defined as a structural design of an interface (Cho et al., 2009; Kim, 2009) and teaching interaction procedure is a systematic form of teaching used by teachers to gauge behaviour (Leaf et al., 2010). The important aspect of the resource for teachers is taking care of the children, technology is a method to help children. The first thing we want to promote is that the content is easy to create and flexible for the teachers, providing positive motivation to join the activities.

Interface design exists at the junction of computing sciences, design arts and social sciences. Human-computer interaction (HCI) is a discipline concerned with the design, evaluation and implementation of interactive computing systems for human use (Rosinski et al., 2009). The purpose of interactive design in HCI is to improve the experience for students with disabilities of direct feedback. Computer-mediated communication facilitates the understanding of communication patterns, forms, functions and subtexts, which can in turn engender an understanding of how to derive meanings within such contexts (Bower & Hedberg, 2010). Patterns or figures are increasingly being used not just in education, but also in many other areas such as software engineering, engineering and business management, and are also frequently advocated for teaching HCI principles (Kotze´ et al., 2008). Interactive whiteboard systems comprise a computer linked to a projector and a large touch-sensitive electronic board displaying the projected image. Children or teachers can operate the content from projector directly by an IR device (Warwick et al., 2010). Because intuition is an operating feature of assistive technology, it is not necessary for operators to use a tool, such as a mouse or keyboard, to learn how to control it. Therefore, it is easier for users to touch the low-cost assistive technology. In this way, information and communication technology is a powerful tool for learning and rehabilitation, which has prominent influences on helping teachers explain difficult concepts, giving access to a huge range of examples, resources and inducing pupils to engage in learning easily (Waite et al., 2007).

This chapter focuses on making some interactive devices at a low price to help children. In studies related to the interactive whiteboard, it is indicated that the interactive whiteboard technology has the potential of supporting teaching and learning (Tataroglu & Erduran, 2010). This assistive technology of interactive design describes a new design for teaching materials developed in the frame of a research project supported by information interface tools (Lin et al., 2010). The introduction of basic education or special education-based computer-aided tools in the routine development process of education is truly important. The display of multimedia teaching materials relies on the operation of a mouse; however, for students with disabilities, a PC mouse is not a good tool, because it is a load for children in the process of learning.

One solution was to explore the application of devices used in contemporary gaming technology, such as the Nintendo Wii or AirMouse (López, 2010). An IR camera was generally used in tracking systems, which lead often to unaffordable costs; particularly, Wii remotes used as IR cameras (Shih, 2011). Some research combined a Wii hand controller and IR emitter to create a low-cost interactive whiteboard, so that teachers were enabled to design teaching materials that enhanced learning for children with developmental disabilities. The interactive technology consisted of a Wii hand controller, IR emitter, a laptop and a projector. This kind of customized design is low-cost at less than US$35. Combined with low-cost materials to create a cheap device, the laptop could be controlled by the wireless IR emitter device that functioned much like a PC mouse (Lee, 2008; Lopez, 2010). The application of an IR emitter is similar to a normal screen update to touch screen.

The application of a low-cost AirMouse through the Wii hand controller was similar to putting a PC mouse in the participant's hand; the design of the low-cost AirMouse and virtual interfaces adopted an interactive method, while the design of teaching materials adopted Flash software that could invent interesting displays in order to raise children's curiosity and increase the rate of use (Lin et al., 2011).

Recent advances in assistive technology have resulted in the mass production of the low-cost Wii hand controllers, which incorporate a small IR camera capable of tracking light sources and transferring their pixel position via Bluetooth to a computer (Clark et al., 2011). Using Wii as a tool in the teaching carried out in the practical results of the promotion has been very effective (Amici et al., 2010). Interactive technology to promote the ability of understanding via the teaching materials, links the experience of learning styles (Lin et al., 2011).

The design of learning interfaces adopts an intuitive design, while the design of teaching materials adopts Flash software to invent interesting animations in order to raise students' learning motives. In addition, students can not only truly experience the vivid teaching materials, but are also impressed by them.

This chapter also tried to develop a low-cost and doable way to make an AirMouse and interactive virtual interface, especially for people wishing to utilize a low-cost device to help children with disabilities.

The virtual interface design had an advantage in that the application was able to make corrections to the teaching materials, while the assistive technology may also be transferred to other training courses. However, through the low-cost AirMouse, the process became uncomplicated for the children; they could operate the tool by intuition. This chapter focused on the equipment in helping children with disabilities and tried to enhance the quality of the equipment which could be tailored to people's different needs.

Some of the related research focuses on a Wii balance board to help people to train their activities (Bateni, 2011; Young et al., 2011). In this chapter, we also discuss the issue of rehabilitation. With rehabilitative work users do it by their willpower, but do not use feedback, in fact, the process of rehabilitation is so boring, people do not have the motivation to do the work. Especially for children with cerebral palsy and the repeated action needed to train for physical activity [c1]. Cerebral palsy is one kind of series of obstacles caused by an immature brain cell injury, including psychological and social adjustment, active function, walking function and daily living etc. Because the plasticity of the brain is high, with the relevant medical rehabilitation for children with cerebral palsy as soon as possible, early training and learning could promote the ability of children (Valencia, 2010). The final cases of this chapter make an application of the Wii hand controller to help children with cerebral palsy using interactive work during the process of rehabilitation. Children with cerebral palsy might not have the ability to control the computer well, but custom-made alternative devices are always more expensive; one kind of solution is to explore the application of devices used in contemporary gaming technology, such as the Wii (López, 2010) or Xbox (Xynos et al., 2010). An IR camera is generally used in tracking systems and often leads to unaffordable costs, particular the Wii hand controller which is used as an IR camera (Shih, 2011) so that teachers are enabled to design interactive custom-made materials for children with cerebral palsy, enhancing learning interest for children with developmental disabilities.

Through the Wii hand controller and IR emitter, the process becomes uncomplicated for the children. Although traditional interactive teaching materials may be suit normal children, there is some burden for children with different disabilities; this chapter focuses on using the advantageous parts of the interactive effect and on creating a cheap device to help children with disabilities to enhance and improve their activities.

2. Method

Interactive whiteboard to promote the ability of understanding in the teaching materials links the experience of learning styles (Lin et al., 2011). Children with disabilities were unable to use base computer control devices, but custom-made alternative devices were always more expensive. Thanks to Johnny Chung Lee (Lee, 2008) a program about low-cost multi-point interactive whiteboards using the Wii hand controller was created and shared freely on his website tp://johnnylee.net/projects/Wii/. The fore part of a Wii hand controller is equipped with an IR camera which enables receiving IR light, while its interior provides Bluetooth for communication purposes; that is to say, it is available to connect with other computers possessing by Bluetooth apparatus. Therefore, the chapter is able to integrate the outcome of Flash and to develop teaching materials for children.

This chapter is based on the Wii hand controller and attempts to design an application of a low-cost interactive virtual interface. The Wii hand controller is a handheld device just like a television remote, with a high-resolution high speed IR camera and wireless Bluetooth connectivity. When the Wii hand controller added an IR filter lens, the Wii hand controller camera was sensitive only to bright sources of IR light emitter. Moreover, the tracked objects must emit a significant amount of near IR light to be detected. Other IR video cameras can obtain this function, but the Wii hand controller was the cheapest. That met the research's goal which was to make the equipment as cheap as possible so that children with disabilities could be helped in a very easy way.

The extended interaction concept enabled the user to stand in front of the screen to operate using the IR emitter pen. This concept was similar to the AirMouse idea. This design will help some children who have difficulties in using a mouse. The Wii hand controller was Wii's remote controller and it was fixed in this research design. The interactive virtual interface device in this research needed a laptop, Wii hand controller, IR emitter pen and cursor calibration software. Besides the laptop, all the cost would be under US$35. A Wii hand controller camera was sensitive only to bright sources of IR light emitter and tracked objects must emit a significant amount of near IR light to be detected.

It would be hard for the children with disabilities to learn to find a position where the sensor of the Wii hand controller would not be blocked. For children with disabilities, they needed to come close to the screen and operate. In the experiments, the children with disabilities took the IR emitter pen near the screen and they covered the IR emitter easily. That made the Wii hand controller unable to receive the message from the IR emitter successfully, so that interactive activities could not work smoothly. It would be easy for normal children to be asked to hold the IR emitter pen as a PC mouse using a hand, and at the same time, pay attention to not blocking the Wii hand controller and the IR emitter pen transmission direction. However, for children with disabilities or even multi-developmental disabilities, they have more difficulties in the learning process of using an IR emitter pen.

In this chapter, the Wii hand controller-based, interactive design build interesting applications. The Wii hand controller makes a simple interactive device and in recent years many people and researchers in the relevant studies have opened up their programme codes free of charge to download [c13]. For the purpose of this chapter to be applied to help more children with disabilities, the price of equipment is the key consideration.

The device in this chapter is easy to use; what participants have to do is to operate the IR light, using a prototype which relies on designing to the specification of users. In addition, the apparatus can adjust the size and scope of projection.

2.1 Low-cost interactive whiteboard

The theory that this chapter applies is making use of Bluetooth to connect a computer and a Wii hand controller. By means of an IR emitter, a Wii hand controller can track the accurate launching location of IR light and then generate a simple whiteboard. According to the course design of participating teachers, image projection based on low-cost interactive whiteboard is divided into two categories.

The first category: a projector shows images on the wall or the projector screen, letting the Wii hand controller point to the projection position. Based on the requirement of students, the range of exhibition can be adjusted. For students with extreme short sight, for instance, we can make use of amplification displaying a larger size in order to improve inconvenience for short-sighted students because the presentation on normal computer monitors is too small, as Fig. 1 illustrates. The second category: letting the Wii hand controller point to a computer monitor which originally merely has the display function. With the assistance of Wii hand controller, the computer monitor thus becomes a touch screen, as Fig. 2 illustrates.

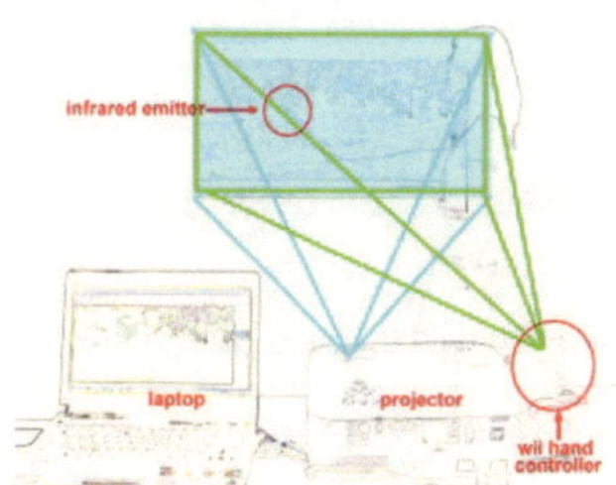

Fig. 1. How to use the low-cost interactive whiteboard on a wall

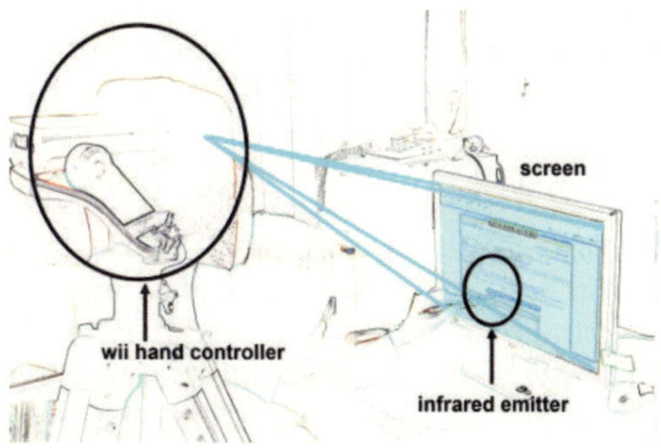

Fig. 2. How to upgrade a screen to a touch screen

The display mode of these two methods is an intuitive learning tool. The operator does not need to use a mouse as a tool, which is especially useful for children with learning disabilities who therefore do not have to learn the use-pattern of the PC mouse. Students will raise their interest towards learning and lessen the generation of frustration during their learning processes.

An interactive whiteboard is an interactive display interface that connects to a computer and a projector. A projector projects onto a board's surface where children can control the display using a pen or other devices, but it is very expensive so teachers cannot use them in many places. Since the Wii hand controller can track sources of IR light, it can track IR light from theled in the tip. In this research, teaching materials are designed using Flash software and the interactive whiteboard is used to develop children's abilities. According to the requirements and preferences of students, participating junior and elementary teachers devised and adjusted the design of the IR emitter.

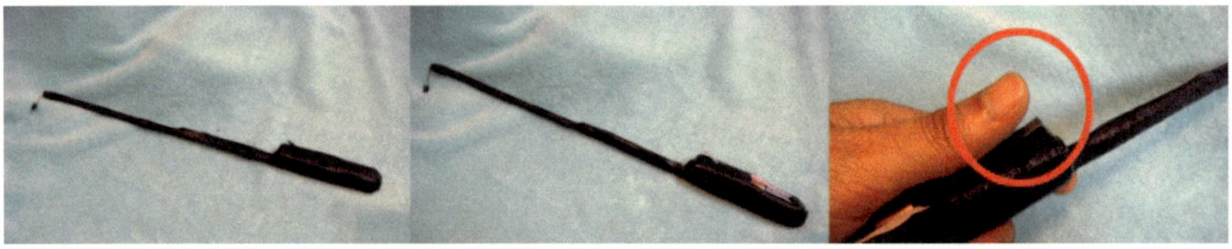

Fig. 3. IR emitter design 1 (demonstration by Te-Hsiung Chen)

Fig. 3. is designed by teacher Te-Hsiung Chen. Because children with disabilities are generally lack the ability to manipulate operational tools and their action sensibilities are also not nimble enough, teacher Chen therefore makes use of micro switch to launch the IR emitter. In addition, with regard to the IR emitter, in order to prevent the Wii hand controller receiver from being interrupted by the location of the body, teacher Chen extends the launching section. This improved design creates distance between the hand and the launching section, and the effect is better than before.

2.2 Application of the low-cost AirMouse for children with physical disabilities

The AirMouse is a device for people to control a computer. The motion-sense technology inside the AirMouse delivers precise in-air cursor control and convenience. The principle is through motion-sense technology which translates your hand movements into on screen cursor motion for in-air operation. The AirMouse not only works on the desk, it also can work in the air or anywhere else. It is more comfortable for users to use and compact and portable enough to be easily used, using natural and comfortable wrist movements to take in-air control of your computer.

This part will show how to use the Wii hand controller and IR emitter not only to make a low-cost interactive whiteboard, but also a low-cost AirMouse. Just like many videos shown on YouTube, the low-cost interactive whiteboard was good for children to use, but the light of the projector would be a burden for people who used it. It would be good for a teacher to start the cursor calibration software, fix the IR emitter pen position and tracking utilization for children. If tracking utilization was lower than 50%, the Wii hand controller position should be adjusted and relocated. By doing so, people could upgrade their normal screen to a touch screen. The participant could sit in a chair, holding an IR emitter pen as a PC mouse in the learning process. It is easy to make a virtual interface; forgetting the method of how to make a low-cost interactive whiteboard, face the IR receive of the Wii hand controller toward the user and prepare an IR emitter.

Just as Fig. 3 shows, follow these steps to set up the virtual interface. Step 1: double click the executable file of Wiimote or Wiimote smoothboard than make sure you go into the situation of "Quick calibration" (Wiimote button A). Step 2: you can see the first red point of calibration, don't let your IR emitter touch the screen, just think of a virtual interface behind you, and push the micro switch of the IR emitter at the corresponding position. Step3: when the second red point of calibration shows on the screen, just like step 2, think of a virtual interface again, and push the micro switch of the IR emitter at the corresponding position. Step 4: follow step 3 and finish the four red points; when the four red marks have been calibrated, you have created a virtual interface. So, the users do not approach the screen or the wall that the projector projects upon, the user is only sitting on the chair, using the IR pen like an AirMouse. The Wii emitter needs to face the user.

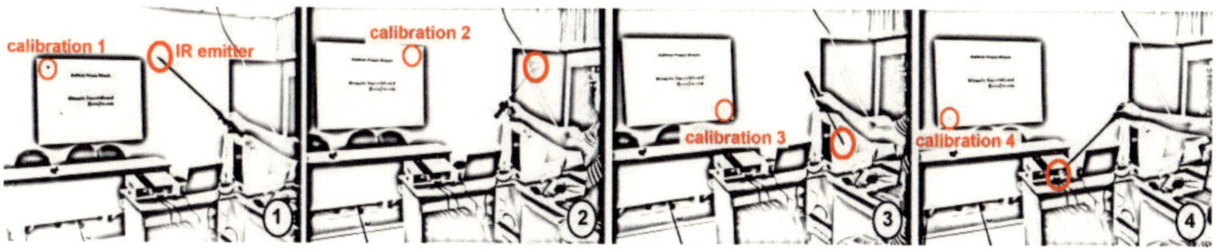

Fig. 4. How to calibrate the virtual interface

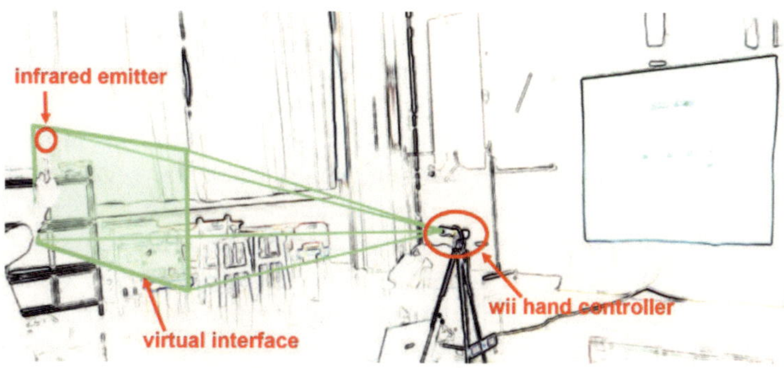

Fig. 5. Low-cost AirMouse and virtual display

The purpose of this chapter was to help children with disabilities to enjoy their learning process. Assistive technology was a helpful method for learning, which had prominent influences on helping teachers explain difficult concepts and giving access to a huge range of examples and resources. This low-cost AirMouse could also help children with disabilities to learn easier; the principle is illustrated in Fig. 5.

This chapter benefited greatly from many experts who devoted their interactive whiteboards technology using the Wii hand controller. Using the IR emitter and Wii hand controller an AirMouse-like system was created - a low-cost and custom-made tool. Therefore, it was able to integrate the real-time feedback via Flash software and PowerPoint, and also contribute to developing teaching materials for children with disabilities.

The theory was applied using Bluetooth to connect the computer, the Wii hand controller and the IR emitter. Through the IR emitter, the Wii hand controller could track the location of the IR emitter. The function of the IR pen was just like an AirMouse, and the children with disabilities could use it easier with a pen instead of an AirMouse as a controller. In order to suit the needs of children, the weight of the device held was only that of a simple pen with a micro switch. The display mode of this method is an intuitive learning tool and better than an AirMouse; children with disabilities would raise the sense of achievement. In virtual display, whenever the children pressed the micro switch in this area, the IR emitter could control a system just like an AirMouse. The virtual display showed the function of the IR receiver. It is like a pen for children to control via the micro switch, and the teaching material could be modified for other courses.

It is a simple tool for children to control the micro switch and the teaching material could be modified for other courses. In the research, the researcher used Flash software to design interactive teaching materials. When the researcher set up the device in the classroom, only the children with developmental disabilities who took an IR pen behind the virtual interface could make use of the interactive effect. Participants from the resource classroom could recognize how to operate this virtual interface, because it only took one object just like a pen which could be waved to see the content.

2.3 Application of low-cost interactive rehabilitation

An IR light array and reflected band could make a low-cost interactive rehabilitation tool.

Reflected band is one kind of flexible tape, it could be tied on the different part of children's body then do to rehabilitation .A projector projects onto a wall where children can control the display using the relative devices, but is very expensive and not portable, so it is not convenient for teachers to use the device in different places. By means of an IR light array and Wii hand controller via reflected band, it could track the relative position of the reflected band and then became an interactive teaching material. A projector shows images on the wall, letting the reflected band reflect from the IR light array to transfer information to the camera sensor of the Wii hand controller. The weight of the device which is held by the children is only a few grams, which is sufficient for the children's requirements. The display method of this part is an intuitive learning tool. The operator does not need to use a PC mouse as a tool and the children will raise their interest and lessen their frustration during the learning process. Since the Wii hand controller could track a source of IR light array, as Fig. 6 shows, so the research could apply this technology to make a low-cost device. A reflected band on the leg or other parts of the body are good examples that could benefit from minimizing tracking instrumentation, as Fig. 7 illustrates.

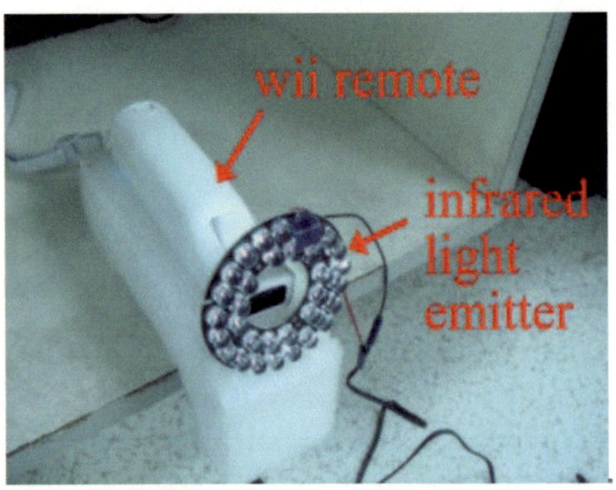

Fig. 6. Simple IR emitter

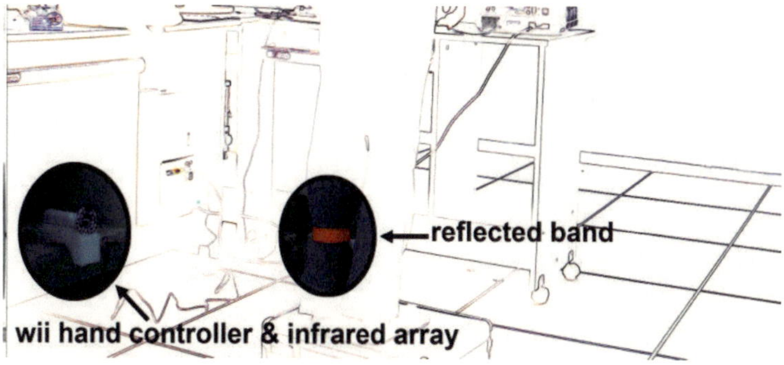

Fig. 7. Simple IR emitter

The Wii hand controller is the Wii's control, from the training process, this chapter illustrates making the Wii hand controller point to IR emitter by which it can detect the IR emitter signal. If the participant achieved the action, the screen will show the contents from PowerPoint next page. The interactive design needs a laptop, a Wii hand controller, an IR emitter, all costing under US$ 35. As teachers design teaching materials, they need to consider the funding constraints which prevent providing interactive teaching materials. Therefore, using a Wii hand controller as a connecting tool, the normal projected screen could upgrade to an interactive device. For teachers to prepare teaching materials depends not only on low-cost, but also easy ways to learn, and an effective interactive interface is a good way to help this process. In the resource classroom or special classroom, teachers care for children with different disabilities, and particularly children with cerebral palsy need physical rehabilitation.

We modified the method connected with the Wii hand controller and IR emitter. With the IR emitter we simplified the emitter bulb and button cell battery, and the weight is no burden for children who need to exercise different parts of muscles according to their different situation. It is easy to process this modification, as shown in Fig. 8.

Fig. 8. Simple IR emitter

The theory applied making use of Bluetooth to connect to the computer, the Wii hand controller and the IR emitter. Via the IR emitter, the Wii hand controller could track the location of the IR emitter, so, the IR pen functioned just like a PC mouse. The children with disabilities could use the pen easier than a mouse controller. Based on the requirements for the children, the weight of the device held is only that of a simple pen with a micro switch. The display mode of this method is an intuitive learning tool. Children with disabilities will raise the sense of achievement. Since the Wii hand controller can track sources of IR light emitter, so the research makes a virtual interface and there is a virtual operating range. When the children press the micro switch in this area the IR emitter acts just like a mouse to control to system, as shown in Fig. 9.

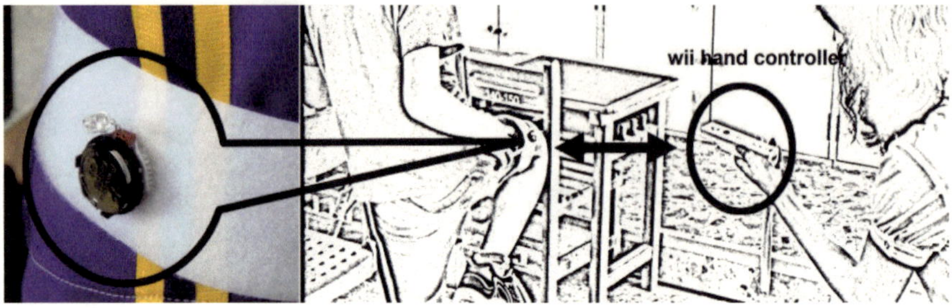

Fig. 9. Simple IR emitter

2.4 Participants

All the participants in this chapter are Taiwanese children. This chapter focuses on two parts. Firstly the design of the digitized teaching materials is examined and then when we experimented in the lab and test it inor resourse classelementary, we revised the ideas on the low-cost interactive whiteboard so that teachers could better participate in the research. The first step is to select teachers from elementary and junior high schools, and make sure they have experience in basic computer skills. Then we discussed how to design the contents for children with disabilities. The teachers designed interactive teaching materials, according to the needs and ages of the students they teach in their schools. The second step focused on how to make the relevant device to suit their children. All the children who participated were from Taiwanese elementary and kindergarten schools.

3. Case study

Children always need to repeat training and students with learning disabilities have difficulty concentrating on teaching materials; this chapter therefore attempts to introduce the concept of custom-made learning for teaching materials. The learning materials are customized for students themselves so as to improve their learning interest. The range of custom-made applications is considerably extensive, including attire, architecture, medical care, rehabilitative instrument etc. This chapter begins from assistive teaching, which provides diverse learning methods in the design of teaching materials. Using an IR emitter and interactive whiteboard can link to corresponding information, which is able to increase the attraction and intimacy of teaching materials. Here are six cases and explanations of this.

3.1 Case study 1 (based on a low-cost interactive whiteboard)

This teaching material could be used as an intuitive class resource. The teaching materials made using PowerPoint, so there is no burden for teachers. In this teaching and learning activity, the teacher introduces and demonstrates how to operate first, and then the student operates it on his/her own, but in this way, it is also easy in a normal class where the children could handle a PC mouse without any working load. Fig. 10 shows the teaching materials in Chinese. In Fig.10 the Chinese words mean MRT(Mass Rapid Transit) and the teacher demonstrates how to put the IR emitter appliance on his finger. When he touches the wall, the macro switch will let the IR emitter work, so the Chinese word can be moved. This design of digital teaching material offers an intuitive way of learning for students when they learn new words. Teacher Huang considers children are not good at holding a pen, so putting the IR emitter on just the ring finger is easier and furthermore, teacher Huang has also devised an IR emitter in glove form.

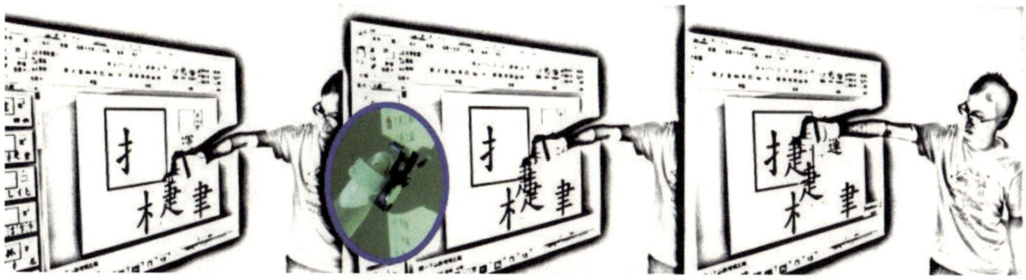

Fig. 10. Application of Chinese teaching materials designed by Kenendy Huang

The tool is in a start up state when the child's palm presses on the wall, which seems as though the machine responds once their hands touch the wall. It has certainly humanized the design for children to absorb knowledge because it is not necessary to teach them how to operate learning tools.

3.2 Case study 2 (based on low-cost interactive whiteboard)

Case 2 is the teaching of Chinese characters. The teaching material of Chinese characters is divided into categories, just like stationery, fruit, transportation, sport, furniture, animals and kitchen appliances, respectively, that are basic words for children with disabilities. In the content of Fig. 11 there are the nine Chinese words within the category of stationery, including "pencil", "crayon" , "ruler", "textbook", "exercises", "school bag", "scissor", "glue" and "tape". Therefore, the teacher used a multi-media teaching method of Chinese characters, which is different from the traditional style of learning, and used pictures and sounds to help him memorize phrases. When clicking on a particular word, a picture and sound will appear accordingly.

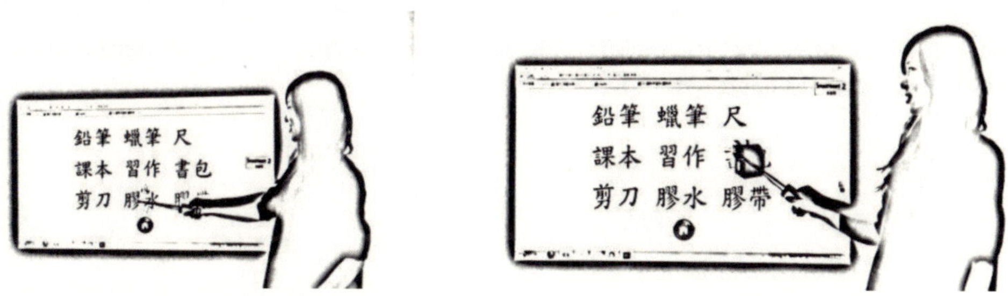

Fig. 11. Course of space training

Particularly in the lesson of recognizing words, the teacher recorded the pronunciation of the corresponding word and used the connection between the word, the pronunciation and the picture, so that the children could receive indirect feedback when only touching the wall. The teacher makes use of the good interactive speciality of the Wii hand controller to design this lesson. The teacher is able to understand students' cognition of left and right, and concepts of extensity. Students show that it is very novel to use a Wii hand controller, as shown in Fig. 11.

3.3 Case study 3 (application on a low-cost AirMouse)

Many children didn't control their posture when using a PC mouse with the participants leaning toward the screen closely. Another problem is that the PC mouse should be used on a platform like a desk, but with the image projected on the wall, they are not on the same axis, so the PC mouse is not suitable for children for disabilities. Now by the advance of technology, the AirMouse could be used in the air to control the computer, it could be through try-and-error process before use regular AirMouse, that will reduce their motivation and increase their burden for children with disabilitiesbut it maybe suit for normal student but for students with disabilities, it could be take more trainings to use it, for children with disabilities ,it will reduce their motivation and increase their burden. This case suggested that a PC mouse and a regular AirMouse were not convenient to use for some children with disabilities. Therefore we developed other equipment for these children, so that they could interact more easily in the learning process. Thus, we designed a low-cost interactive interface to act as a substitute for a PC mouse.

The participant was a child studied at resource class. With the use of this low-cost AirMouse, it could train the ability of hand-eye coordination. When the child operated the low-cost AirMouse in the virtual interface range, the screen showed the relative image. The do-it-yourself AirMouse was very easy to use. There was a micro switch on the IR emitter pen, when the child pressed on it, it came on. On the other hand, when the child did not press on the switch, it stayed off. The size of the virtual interface is adjustable, so depending on the participant's hand-eye coordination, the appropriate virtual interface size could be made for different users. The content was designed using Flash software - teachers could easily change the content via updating different pictures. It is no burden for teachers to prepare the teaching materials thereby supplying more motivation to use this low-cost AirMouse to help their children. Teachers could use this device to train their students' activities, as illustrated in Fig. 12. In addition, the operating method of this low-cost AirMouse is easier than the regular AirMouse; the method allows the child to hold in the style of a pen and press the micro switch then wave, as such the child could erase the blank area and see more parts of the picture. This way, teachers could change different pictures to attract children to do hand-eye coordination training.

Fig. 12. A participant's posture when operating a PC mouse

In the research process, the research demonstrated how to use the assistive tool to help the child with disabilities, and the child was asked to hold the IR emitter pen in her hand. Then the child used the IR emitter pen to do work shown by the researcher. The child only needed to hold the IR pen, press the micro switch, move the pen in the air in any direction, thereby the child could see the movement of the low-cost AirMouse reflected on the screen. In this case, the participant sat on a chair and held a low-cost AirMouse in the virtual interface range, when the child pressed the micro switch of the low-cost AirMouse, it was just like she was operating a real AirMouse.

In the teaching of using this low-cost AirMouse, the child learned how to use it very quickly since it was similar to how she used a normal pen. After observing the child in this research, the researcher found that using this low-cost AirMouse allowed the child to interact with the computer learning programme with appropriate gestures, additionally, it allow the training of hand-eye coordination.

3.4 Case study 4 (application on a low-cost AirMouse)

These cases are custom-made for children with disabilities in kindergarten to increase their attention. Using a virtual interface and low-cost AirMouse could make an interactive design, which is able to increase the attention on the contents of teaching materials. This case demonstrates how we use interactive effects to help a child who is five years old in kindergarten. In this case the researcher showed how to use the low-cost assistive tool and then the young child was asked only to hold it in the style of a pen to control the computer. When he pressed the micro switch of the pen, it was just as if using an AirMouse and the child could see the images that the projector projected on the wall.

The teaching materials in this case were custom-made by Chang Ling-Wei, using the low-cost AirMouse to help her to communicate with the child. Because the child has developmental disabilities and did not want to talk, she used this device to attract his attention. She show the slides page by page and asked for the child follow her in reading the relevant words and doing the relevant actions - when he finished, he could use the low-cost AirMouse. The child did not talk during his class or even at home, he just said a few words when doing this activity.

The case is not only a custom-made teaching materials design, but also a real-time feedback. Fig. 13 shows the teaching materials. Because their curriculum is focused on how to brush teeth the care about themselves, a volunteer, Chen, made the relative teaching materials as assistive materials to help him. This case focused on how using low-cost assistive technology could have an interactive effect; children are offered a digital presentation of designing and learning concepts for easier ways to operate, in this way, children can train their body ability by virtual interface design with interaction.

In the experience of many experiments, the children with disabilities took the IR emitter pen near the screen, they covered the IR emitter, making the Wii hand controller not able to successfully receive the IR emitter of the message, so that interactive activities could not work smoothly.

Normal children were asked to hand hold an IR emitter pen as a PC mouse and at the same time pay attention to not covering the Wii hand controller and the IR emitter pen transmission direction. For normal children's understanding it should be a simple command and also very easy to implement, but for children with disabilities or children with multi-developmental disabilities, the learning process could potentially cause a lot of trouble with the learning load.

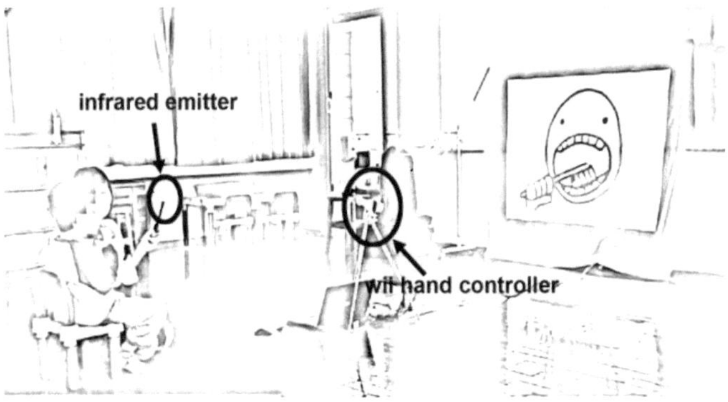

Fig. 13. A boy participant in a custom-made course

3.5 Case study 5 (application for rehabilitation)

The teacher makes use of the good interactive speciality to design this lesson. The exercise of step test is to step up with one foot and then step down via a box, the height of this box is approximately 10 cm. This is a basic exercise for children to train their bodies. In this case, the researcher used a tool box with a height of almost 10 cm, then put a reflected band on one of the participant's legs. There are four steps in the exercise, children must follow the steps when asked. Step 1 lifts the left foot on the tool box, step 2 lifts the right foot on the tool box, step 3 puts the left foot down from the tool box, step 4 puts the right foot down from the tool box, the projector will project the relevant number to encourage the participant. So, in case 5, the participant could learn two skills at one time, one is the skill of counting and the other is training his physically abilities - it is a win-win strategy, as shown in Fig. 14.

Fig. 14. A participant's tie a reflected band as posture of operatinga pc mouse

The strategy of case 5 is to design different physical activities to help children with disabilities and help them enjoy their rehabilitation process. Through the use of assistive technology participants have a positive attitude toward rehabilitation. The contents are only designed through the use of PowerPoint. Teachers or parents could make the contents more interesting than this case, preferably the contents could be connected to the participant's life to attract the users to do the routine rehabilitation. This case 5 was appreciated by many experts who devote their technology to interactive technology using the Wii hand controller. With IR array, reflected band and Wii hand controller the contents could be custom-made and flexible. Therefore, case 5 is able to integrate the real-time feedback of PowerPoint to develop different teaching materials for children with disabilities.

3.6 Case study 6 (application for rehabilitation)

Case 6 is not only focused on a low-cost device, but also paid attention to easy design and application. This case aimed at promoting and redesigning the process of interactive rehabilitation, so we used the simplest method to do the process of rehabilitation. Because being low-cost and easy to follow are very important factors for teachers and parents, low-cost assistive technology could make interactive effective. Children with disabilities are offered digital presentations of designing and learning concepts for easier ways to operate. Children have cerebral palsy due to different injuries and many often have other barriers, including visual impairment, hearing impairment, speech impairment, mental retardation, learning disorders and so on. For development and life, children with cerebral palsy need a lot of help and practice, but they lack motivation and also a lack spontaneous interaction with the external environment. Children can do the rehabilitation for their coordination. The participant in this case is a child with cerebral palsy in elementary school, she must do rehabilitation everyday to keep the abilities of her hand-eye coordination. We hope to keep the basic body motion using digital technology. Daily and routine rehabilitation for training the energy of her body is her basic work, but there is no immediately feedback from traditional rehabilitation and the child found it difficult to give full effort to her exercises. Case 6 wanted to use a low-cost and do-it-yourself interactive tool to enhance the rate of her daily rehabilitation. In case 6, we only put an IR light with one button cell to make sure the IR emitter is at the working position. Because the button cell is a small single cell battery shaped as a squat cylinder like a button on a garment, they are used to power small portable electronics devices. Therefore, it has enough power for IR light in the working state, so we can control the weight as light as we can. We put the low-cost and do-it-yourself device on her wrist, when she put her hand up, the Wii hand controller received the signal of the IR emitter and gave an instruction via Bluetooth to let the contents of PowerPoint jump to the next page. Because the teaching materials are custom-made for the child, case 6 designed the interactive contents from her pictures, therefore, the contents will attract her and the child will have strong motivation to see the next page. She made efforts to raise her hand high again to let the PowerPoint go to next page, it is a special experiment for us to make this light device to help a child under no weight burden, as shown in Fig. 15.

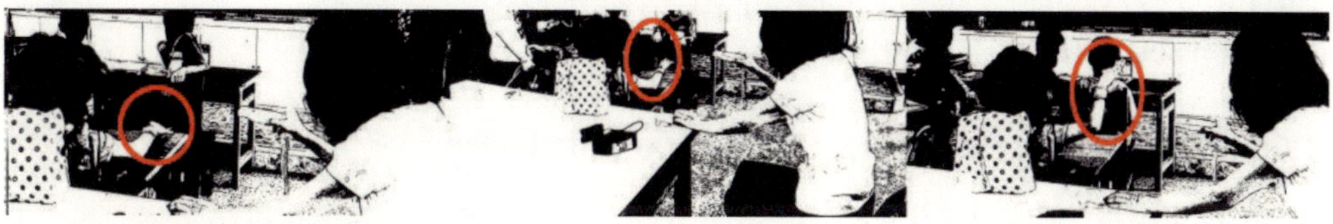

Fig. 15. The experimental process

In the whole experimental process, step 1 connected the Wii hand controller and the laptop, step 2 let the child put on the low-cost and do-it-yourself device, wherever that she wanted to do the rehabilitation. Step 3 made sure the Wii hand controller was toward the direction of the IR emitter and made sure the mode of PowerPoint was in play state. When that was complete the child could do the real-time interactive activity.

Extension of the concept of interaction allows the operator to wear with IR emitter of the tool, depending on which parts of child's body needs to be trained. The IR emitter could be affixed to their hat, wrist, knee etc. The concept of the IR emitter is as the left button of mouse and for children with cerebral palsy it can work different areas of the physical training. With regard to software, teachers only designed teaching materials using PowerPoint. For teachers, because it is the basic software used in class, teachers do not require additional software training, making it a better way for teachers prepare the content of teaching materials for children and as such they could pay more attention and motivation on the training process for children with cerebral palsy.

In case 6, the child was very happy to finish the activities and completely finished her daily rehabilitative work for her arms. When she finished the repeated practice, she was sweating and felt very tired. Her teacher also participated with case 6 and observed all the procedures; the teacher pointed out that the real-time interactive feedback is an important effect for her, so she paid attention to her rehabilitation and really stretched her arms for full effect.

4. Conclusion

The chapter used case studies to discuss low-cost interactive devices to help children with different disabilities. From the processes of this chapter, the obvious differences between the traditional and the custom-made design of teaching materials could be observed when children needed different help. Children also enjoyed the custom-made and custom-teaching materials. Therefore, according to the outcome of the six cases, the goals of the research were achieved. We focused on using cheap devices with assistive technology to create directly interactive effects; children were offered digital presentations of designing and learning concepts for easier ways to operate. In this way, children could improve their gestures when using a computer as an interface.

Using assistive technology as a tool to enhance the variety of teaching materials is a trend. So, the assistance of the Wii hand controller and IR emitter could support teachers and students with disabilities giving more chance to improve their situation. Particularly for children with disabilities, they can obtain real-time feedback via low-cost devices that increases their motivation toward learning processes or rehabilitation. Our conclusion is divided into three parts. The first part is a conclusion about the low-cost whiteboard, the second part is conclusion about the application of the low-cost AirMouse, the last part is a conclusion about the application for rehabilitation.

4.1 Conclusion about the low-cost interactive whiteboard

For teachers who work with children with disabilities, designing teaching materials and interface design of the learning process could be user-friendly. Some skills are very spectacular, but not easy to understand and one must have a programming background. We found it was too difficult for teachers to learn so the purpose of our topic was easy learning that will create interest for teachers and their students. Some of the topics in designed teaching units make interactive interface design of assistive technology as their learning goals. Technology-based learning focuses on content learning in order to explore more user-friendly and collaborative approaches in their active learning aspect.

From case 1 and 2, the children concentrated on the interactive content during the operating process. Most of the children were very interested in digital teaching materials. It was found using multi-media to teach and learn is easier and gives rise to greater student motivation than general activities which use books or pencils and papers. With pictures and sounds as hints, the scaffold of learning is built up. From the children who used the interactive teaching materials, some children felt that the learning process was no longer so difficult.

This multi-media teaching material is presented through multi-senses, which makes the content more vivid and adds interest to the process of learning. Students can also choose learning methods (visual approach, auditory approach) which is suitable for them to learn, this creates a deeper impression and the effect of learning is better; moreover, it saves time. Furthermore, thanks to the application of an information interface, students are inclined to absorb knowledge actively and aggressively. They can learn independently with no need of assistance. The interactive media interface for children with learning disabilities has become an important and helpful computer-aided design for teaching materials. The interface arrangement also gives students assistance when they attempt to learn different units. There is a close link between assistive technology, special education and communication design research, as well as studies that examine how interactive design of teaching materials can influence learning. HCI is a valuable issue for disabled students' learning in the future. By using the Wii hand controller to develop an interactive whiteboard, this made it cost-low and easy to carry. It was also good for the classroom with a limited budget. However, the eyes of the user could not avoid the light of the projector. Although people could use a short-focus projection machine to solve the problem, for teachers who needed to teach from one school to another, it would be hard for them to carry the equipment. This difficulty was a common problem for people who used a low-cost interactive whiteboard or normal interactive whiteboard. This chapter focused on children with disabilities and tried to upgrade a normal screen into an interactive whiteboard using a Wii hand controller. This only cost under US$3540. In this chapter, the researchers made the Wii hand controller face the user and the user did not need to point the IR emitter pen close to the screen or the wall. By doing so, the user did not need to face the light of the projector; in addition, it could correct the user's inappropriate gestures.

4.2 Conclusion on the application of the low-cost AirMouse

There could be more applications in the field of virtual interface because the virtual interface focused on easy operation and easy use, moreover, the virtual interface provided not only real-time feedback, but also a lower price to execute the experiment. It was real and helpful to design interactive teaching materials for teachers, especially for resource and supply teachers. Based on the fundamentals of making the operating interface simpler and burden-free, the assistive technology used by elementary school students and kindergarten children to produce fun learning.

The main theme of the project was the assessment of the application of a virtual interface for the design of interactive teaching materials; Flash was used as the main application software to produce the interactive contents such as videos and animations. The main concern was to induce the interest of the elementary teachers who have participated in the first stage to use the easy-to-learn software. Since they were already equipped with basic skills to use the software while formulating the teaching materials, the teachers would be able to focus their energy on designing the teaching materials. In terms of the hardware, the IR emitter was modified so that the projected images were interactive in real-time, relieving the burden of using keyboard and mouse for the children. Furthermore, the introduction of interactive, assistive technology also facilitated the children's learning process.

When teaching children from resource classes, the teaching materials were modified according to their needs. The development of case 3 and 4 may cater to children with more diverse needs; by introducing Wii's interactive technology into the project, the project was not only academically sound, innovative and flexible, it also had significant influence on the design and development of teaching materials. During the children's learning process, instead of employing creed-oriented teaching, knowledge was obtained via interaction, a process less onerous for children. The development of digitization possessed high potential; as far as children were concerned, not only was the application of digital content refreshing, since the teaching materials emphasized methods such as interaction and coordination, the children also became more interested while learning, which in turn generated a sense of accomplishment. Therefore, the development of an interactive interface to assist the children's learning is imperative.

By using the application of low-cost interactive technology, the emphasis was placed on aspects such as ease of learning by resource class teachers, low equipment costs and ease of promotion. The sensor from the IR emitter could transmit the signals to the computer and the screen at the same time. To put it simply, we used an IR emitter pen as a mouse, therefore, for continuing research different teaching materials could be designed. The main purpose was to instruct children in the application of physical activities. This chapter tried to design different courses for children with different disabilities.

This type of AirMouse could be developed as an assistive device for children with different disabilities, plus be used by teachers when developing the design of teaching materials. Particularly for students with special needs, via the application of interactive appliances, they could train themselves in an interesting way.

4.3 Conclusion on the application for rehabilitation

Rehabilitation can reduce some problems associated with cerebral palsy. Case 5 and 6 used a Wii hand controller to assist physical therapy for child with cerebral palsy. In this case, by the arm training to explore rehabilitative activity by interactive design, observe the results about the assistive technology applied for child with cerebral palsy.

Just as with the application of low-cost interactive technology, the emphasis is placed on aspects such as ease of learning by resource class teachers, low equipment costs and ease of promotion. The sensor from the IR emitter could transmit the signals to the computer and the screen at the same time. To put it simply, we used an IR emitter pen as a mouse, therefore the continue researchers could design different teaching materials. The main purpose is to instruct children in the application of physical activities. The chapter tried to design more different course for children with different disabilities.

There will be more application on the field of virtual interface, because the virtual interface focus on easy operate and easy use, moreover ,the virtual interface provide not only real-time feedbacks but also lower price could execute the experiment. It is real and quite assistance to design interactive teaching materials for teacher, special for the resource teacher and itinerant teachers. Based on the fundamentals of making the operating interface simpler and burden-free, the assistive technology is applied on elementary school students and kindergarten children to produce fun learning.

In this chapter, in the application of tools to improve rehabilitative methods for children with cerebral palsy, we found that concentration can be increased. By training it can improve hand-eye coordination and improve short attention span, lack of patience and the problem of low concentration. Children with cerebral palsy need to receive long-term rehabilitation and, as with case 5 and 6, interactive real-time feedback is needed so that children are motivated to participate in the process. The interactive process was designed to link the interactive effect to rehabilitation and to positively support the child.

In support of this, in case 5 and 6 the participant only needed to raise her hands and then see the exchange of images. The child's expression was very excited and the child made great effort to raise her hand high to see the interaction of different images. In particular, the content is closely related to the child, as the dynamic characters in the images are all about her real life so she knew the content was custom-made. The interactive feedback is related to her daily life. This is the cause of motivation for rehabilitation. The application of the simple design of assistive devices and materials to the process of rehabilitation means there is two-way feedback.

4.4 Final conclusion

In this era of advanced technological assistive tools, students with special needs have a more convenient life even in the process of learning or exercises.

The primary direction and the main goal for schools and teachers is the efficiently promote the learning process. Teaching materials as well as teaching methods have been continuously developed and innovated, especially in technological assistive tools.

The importance of information and communication skills in interface design in the future has been asserted to support the development of these skills and tools in schools. Students need to possess the capability to use technology and information. Teachers therefore can integrate information technology into teaching activities.

By changing the application of equipment, the results offered children with disabilities different ways of learning, as well as enhanced their ability to pay attention. Since this chapter defined that the teacher should redesign and create the developments, the teaching materials used Flash and PowerPoint software. This chapter focused on using easy software, because if the software is too complex teachers would lose their momentum. The results obtained in this chapter can be extended to different types of activities, especially for children with cerebral palsy, making it very attractive in rehabilitation programmes.

This chapter focus on As material designed to beusing simple materials and software to make an interactive environment, power point can be execute basically, so, it is easy to share and promote in in the applicative fieldon and very suitable for promotion. The design of teaching materials uses common software which can be used to exchange different content easily; it is a great interactive interface for the physical activity of training. This chapter relates to different age categories of children with different disorders and can be adjusted to expand to the scope of services.

5. Acknowledgments

Special thanks to the teachers who participated in this study - without their help this research could not have been finished. These teachers during the arrangement of courses considered individual students' differences. Besides basic teaching aids, they proposed many ideas and practiced them in class.

- Te-Hsiung Chen (National Tainan School for the Hearing Impaired, 52, Xinyi Rd., Xinhua Township, Tainan County 712, Taiwan, R.O.C.).
- Ho-Hsiu Lin (Tainan Municipal Shengli Elementary School, Tainan, Taiwan).
- Yen-Huai Jen (Department of Early Childhood, TransWorld University, Yunlin, Taiwan).
- Li-Chih Wang, Chang Ling-Wei (Department of Special Education, NUTN, Taiwan).
- Yan-Jin Wu, Mei-Lin Hung (Graduate Institute of Assistive Technology, NUTN, 33, Sec. 2, Shu-Lin St., Tainan 700, Taiwan, R.O.C.).
- Kenendy Huang (Municipal Jinsyue Elementary School, 47, Nanning St, Tainan 700, Taiwan, R.O.C.).
- Shu-Hua Chen (Guei-Nan Elementary School, 171, Mincyuan S. Rd., Gueiren Township, Tainan County 711, Taiwan).
- Chia-Pei Liu (Municipal Jinsyue Elementary School, 47, Nanning St, Tainan 700, Taiwan, R.O.C.).
- Shu-Ying Chou (Tainan Municipal Heshun Elementary School, 5, Lane 178, Sec. 5, Anhe Rd.Tainan, Taiwan, R.O.C.).
- Yu-Ling Liu (Tainan Municipal Haidong Elementary School, Tainan, Taiwan).

In addition, special thanks to teachers of the resources classes. This work was partially supported by the National Science Council, Taiwan, under Grant No. 100-2410-H-024-028-MY2.

6. References

Amici S. De, Sanna A., Lamberti F., Pralio B. (2010). A Wii Remote-based Infrared-optical Tracking System, *Entertainment Computing,* Vol.1,(December 2010),pp. 119-124, ISSN: 1875-9521

Bateni, H. (2011).Changes in balance in older adults based on use of physical therapy vs the Wii Fit gaming system: a preliminary study , *Physiotherapy.*,pp.2-7 (May, 2011). ISSN: 0031-9406

Bower, M., Hedberg, J.G. (2010). A Quantitative Multimodal Discourse Analysis of Teaching and Learning in a Web-conferencing Environment–the Efficacy of Student-centred Learning Designs. *Computers & Education,* Vol. 54 (February, 2010),pp. 462-478, ISSN:0360-1315

Cho, V., Cheng, T.C. E., Lai, W.M.J. (2009). The Role of Perceived User-interface Design in Continued Usage Intention of Self-paced e-learning Tools. *Computers & Education,* Vol.53, (September 2009), pp.216-227, ISSN:0360-1315

Clark, R.A., Paterson, K., Ritchie, C., Blundell, S., Bryant, A.L. (2011). Design and validation of a portable, inexpensive and multi-beam timing light system using the Nintendo Wii hand controllers, *Journal of Science and Medicine in Sport,* Vol.14, (March 2011),pp.177-182, ISSN: 1440-2440

Hung, P.H., Lin C.Y., Lu C.C., Chang Y.Y. (2010). Development and Application of Online Assessment for Experimental Debugging Performance. *The 7th conference of the international test commission*,p.102,Hong Kong, July 19-21, 2010.

Kawate, K., Ohneda, Y., Ohmura, T., Yajima, H., Sugimoto, K., Takakura, Y. (2009). Computed Tomography-based Custom-made Stem for Dysplastic Hips in Japanese Patients. *The Journal of Arthroplasty* ,Vol.24, (January 2009), pp.65-70, ISSN: 0883-5403

Kim, Y.J. (2009). The Effects of Task Complexity on Learner-learner Interaction. *System* ,Vol.37, (June 2009),pp. 254-268, ISSN: 0346-251X

Kotze´, P., Renaud, K., Biljon, J.V. (2008). Don't Do This - Pitfalls in Using Anti-patterns in Teaching Human-computer Interaction Principles. *Computers & Education* ,Vol.50,(April 2008),pp.979-1008, ISSN: 0360-1315

Leaf, J.B., Dotson, W.H., Oppeneheim, M.L., Sheldon, J.B., Sherman, J.A. (2010). The Effectiveness of a Group Teaching Interaction Procedure for Teaching Social Skills to Young Children with a Pervasive Developmental Disorder. *Research in Autism Spectrum Disorders* ,Vol.4, (April-June 2010), pp.186-198, ISSN: 1750-9467

Lee, J.C. (2008). Hacking the Nintendo Wii Remote, *Persvastive Computing*.Vol.7,(July-Sept. 2008),pp.39-45, ISSN: 1536-1268

Lin, C.Y., Hung, P.H., Wang L.C., Lin C.C. 2010. (2010). Reducing cognitive load through virtual environments among hearing-impaired students. *2010 second Pacific-Asia conference on circuit, communication and system*, pp. 183-186,Beijing,China, August 1-2,2010,ISSN9781424479689

Lin C.Y., Lin H.H., Jen Y.H., Wang L.C., Chang L.W. (2011). Interactive technology application program of experience learning for children with developmental disabilities. *Advanced materials research*, Vol.267, (June, 2011), pp.259-264, ISSN: 1022-6680

Lin C.Y., Lin C.C., Chen T.H., Hung M.L., Liu Y.L. (2011). Application IR emitter as interactive interface on teaching material design for children. *Advanced Materials Research*, Vols.233-235,(May 2011), pp.1858-1861, ISSN: 1022-6680

Lin, C.Y., Jen Y.H., Wang L.C., Lin H.H., Chang L.W. (2011). Assessment of the application of Wii remote for the design of interactive teaching materials, *Communications in Computer and Information Science*, Vol. 235, pp.483-490, ISSN: 1865-0929

López, O.S. (2010).The Digital Learning Classroom: Improving English Language Learners' academic success in mathematics and reading using interactive whiteboard technology, *Computers & Education* Vol.54 ,pp.901–915, ISSN: 0360-1315

Rosinski, P., Squire, M., Strange B. (2009). Strange bedfellows: human-computer interaction, interface design, and composition pedagogy. *Computers and Composition* ,Vol.26,(September 2009), pp.149-163, ISSN: 8755-4615

Shih C.H. (2011): Assisting people with attention deficit hyperactivity disorder by actively reducing limb hyperactive behavior with a gyration AirMouse through a controlled environmental stimulation, *Research in Developmental Disabilities* ,Vol.32, (January-February 2011), pp.30-36, ISSN: 0891-4222

Shimizu, H., McDonough, C.S. (2006) Programmed instruction to teach pointing with a computer mouse in preschoolers with developmental disabilities. *Research in Developmental Disabilities*. Vol.27, (March-April 2006), pp.175-189, ISSN: 0891-4222

Tataroglu, B., Erduran, A. (2010) Examining students' attitudes and views towards usage an interactive whiteboard in mathematics lessons. *Procedia- Social and Behavioral Sciences* ,Vol.2, (May 2010), pp. 2533-2538, ISSN: 1877-0428

Valencia, F.G. (2010) Management of hip deformities in cerebral palsy. *Orthopedic Clinics of North America*, Vol. 41,(October 2010), pp. 549-559, ISSN: 0030-5898

Waite, S.J., Wheeler, S., Bromfield, C. (2007).Our flexible friend: the implications of individual differences for information technology teaching. *Computers & Education*, Vol. 48, (January 2007),pp. 80-99, ISSN: 0360-1315

Warwick, P., Mercer, N., Kershner, R., Staarman, J.K. (2010). In the mind and in the technology: the vicarious presence of the teacher in pupil's learning of science in collaborative group activity at the interactive Whiteboard. *Computers & Education* ,Vol.55, (August 2010),pp. 350-362, ISSN: 0360-1315

Xynos, K., S. Harries, I. Sutherland, G. Davies, A. Blyth, (2010).Xbox 360: A digital forensic investigation of the hard disk drive, *Digital Investigation*, Vol.6, (May 2010), pp.104-111, ISSN: 1742-2876

Young, W., Ferguson, S., Brault, S., Craig, C. (2011). Assessing and training standing balance in older adults: A novel approach using the 'Nintendo Wii' Balance Board. *Gait & Posture*, Vol.33, (February 2011),pp. 303-305, ISSN: 0966-6362

2

Disabled Pupils` Use of Assistive ICT in Norwegian Schools

Sylvia Söderström
NTNU Social Research, Department of Diversity and Inclusion
Norway

1. Introduction

The subject of this chapter is disabled pupils` use of assistive information and communication technologies (ICT) in Norwegian primary and secondary schools. The chapter investigates how use, or non-use, of assistive ICT at school influences disabled pupils' opportunities for active participation in an inclusive educational system within the regular school system. This investigation draws on a qualitative pilot study carried out in Norwegian primary and secondary schools in autumn 2010.

After years with separate educational systems for disabled children, the prevailing view today is that such educational segregation should be avoided, and that all children should be educated with their peers in regular schools. These inclusive schools are perceived to be *"the most effective means of combating discriminatory attitudes, creating welcoming communities, building an inclusive society and achieving education for all; moreover, they provide an effective education to the majority of children and improve the efficiency and ultimately the cost-effectiveness of the entire educational system"* (UNESCO, 1994:9). This perspective includes more than valuing the inclusion of all in a common education system; it is also perceives an inclusive education system as a means to develop a society that welcomes diversity. Several countries, including Norway, have committed to this view by signing declarations and conventions that intend to provide every child an acceptable level of education, regardless of the child's individual characteristics, abilities, interests and learning needs (ibid.). Moreover, according to the Norwegian Education Act, every child in Norway has the right to attend their local school, and a regular class or group at this local school (Education Act, 1997).

The goal of this inclusive education policy is to provide for the full participation in environments and activities that are common and positively valued, and to remove arrangements that are devalued and stigmatized - such as special arrangements for disabled pupils (Tøssebro & Lundeby, 2002). However, being included is more than being present. Being included in an ordinary school means, *"Being in an ordinary school with other students, learning the same curriculum, at the same time, in the same classroom, with full acceptance by all in a way which makes the student feel no different from any other student"* (Bailey, 1998:184). This means that inclusion is achieved only when the concept has lost its content, and there is no longer any distinction between "ordinary" and "different." To what extent this "all inclusive" education takes place during an ordinary day at school may vary (Wendelborg, 2010). The question asked and answered in this chapter is to what extent the use of assistive

ICT for disabled pupils in Norwegian compulsory school contributes to the full inclusion of these pupils in regular Norwegian schools.

1.2 The Norwegian context

Norwegian ICT policy aims at securing an information society for all, emphasizing the importance of technology for all, including disabled people. One priority area for ICT policy is the compulsory school (MD, 2005). Authorities point to accessible and usable ICT as a vital means in securing the basic principle of inclusion, equality and full participation for disabled people (MD, 2005; NOU 2001:22, Report No. 17 to the Storting, 2006). According to the Norwegian Education Act every child in Norway has the right to attend their local school, and a regular class or group at this local school (Education Act, 1997). How the individual school is organised has, however, proven to be of vital importance for disabled pupils' possibilities for participation and inclusion (Wendelborg, 2010; Wendelborg & Kvello, 2010; Wendelborg & Tøssebro, 2010). Even when a disabled child attends a regular school and is a member of a regular class he or she may not be included in the school setting. Research suggests that disabled pupils in Norwegian regular schools participate less in activities at school than their non-disabled peers, and they have less access to curriculum activities (Erikson, Welander & Granlund, 2007; Shevlin, Kenny & McNeela, 2002; Wendelborg & Tøssebro, 2010). The public ideology is that when special education is needed this education shall take place in a regular classroom setting together with classroom peers. Nevertheless, there is a documented development of establishing more and more special classes in regular schools, and of using a variety of segregated arrangements for disabled pupils (Tøssebro et al., 2006; Wendelborg, 2010). Resent research shows that 70 percent of disabled pupils in regular schools attend their classes less than 50 percent of the time spent at school (Wendelborg, 2010). This means that in spite of an effort to promote an inclusive school system, exclusion and marginalization of disabled pupils still takes place in Norwegian schools. Viewed against this background my question is *if* -, and *how* -, use of assistive ICT for disabled pupils promote or hamper their inclusion and participation in ordinary classroom settings. What is really assistive ICT, and how do Norwegian pupils access this technology? We will examine this closer in the next paragraph.

2. Provision of assistive technology in Norway

ICT has become a central facilitator in Norwegian inclusion policy. This emphasis on ICT is called the digital inclusion policy. This policy has three main pillars: *digital access* for everyone, *universal design* of all ICT, and *digital skills* e. g. by enhanced use of digital learning resources in education (MD, 2005). Two different strategies are used to promote access to ICT for all. The first strategy, which emphasises the universal design of ICT, has focused on public Web sites, open standards, and open sources (Fossestøl, 2007). The second strategy is a rights-based national assistive technology diffusion system. Every Norwegian county houses an Assistive Technology Centre that provides assistive technologies, free of charge, to people whose ability to function in everyday life is considerably and persistently reduced. These centres provide assistive technology solutions for use at home, school, work, or leisure, and for people of all ages. Common assistive ICT include standard computer devices, ICT tools, and equipment for handling the computer and other communication aids (Hansen, 2007).

Providing and maintaining assistive ICT for disabled pupils requires the involvement of many different professions and services. Professions and services which are involved in this process most often are occupational therapists, assistive technology centres, various centers of expertise, technical suppliers, teachers, school IT consultants, and technical departments of local municipalities, along with the disabled pupils and their parents. These professions and services are required to cooperate in testing, adjusting, educating, implementing, maintaining and upgrading the disabled pupils` assistive ICT. The disabled pupils in the current study were all assigned assistive ICT due to their visual or mobility impairment.

2.1 Assistive ICT for persons with mobility or visual impairments

Assistive technology may be defined as any item, piece of equipment, or product that is applied to secure, increase, maintain or improve functional capabilities (Wielandt, et al., 2006). Assistive ICT for mobility- impaired people usually involves equipment that eliminates the requirement of fine hand and finger motor skills. The assistive ICT used by the mobility-impaired pupil in the current study are different software programs, especially an *eye tracking program* called MyTobii, an *enlarged keyboard* called BigKey, and a joystick. These assistive ICT equipment provided access to the computer for this pupil who had very limited control of her fine motor movements.

The main goal of visual assistive ICT is to provide the best possible sight-enhancement or sight-substitution mechanism. For partially sighted people, these goals mean magnifying the screen display in order to facilitate the performance of visual tasks such as reading texts, selecting menus, responding to system prompts, and navigating between different parts of Web sites. Usually this magnification involves the use of a *screen magnifier software application* that runs as a background task. Such screen magnifiers enables users to enlarge text and graphics across a wide range of levels (Chiang et al., 2005).

Providing ICT access to blind persons involves non-visual alternatives for tasks that are traditionally understood as visual, through the use of assistive ICT that translate the visual interface into either tactile or auditory output, or a combination of the two. The most common assistive ICT for blind people is an electronic *Braille display* that produces a refreshable, line-by-line tactile output on a special keyboard. However, as this tactile output is purely text-based, it is less helpful in translating graphic interfaces. To access graphic interfaces a blind person needs *a screen reader*, a software application that interprets and translates text and graphical displays into auditory output. Overall solutions to visual ICT access problems, however successful, require not only that assistive ICT is well-designed, but also that Web site content and layout are flexible and organised to promote accessibility (Chiang et al., 2005). This sight-enhancing and sight-substituting assistive ICT equipment provided access to the computer for two of the pupils in the study, one of them being visually impaired and the other one being blind.

In addition to examining the disabled pupils` use of these technologies, we will also scrutinise what premises are required for the full utilization of the anticipated possibilities inherent in these technologies. As this chapter will show, this investigation provides deeper insight into the significance of some decisive details, as well as contributing new knowledge to this field. However, before we investigate these circumstances in more detail, I will provide a short overview of (i) relevant research, (ii) theoretical perspectives, and (iii) a methodological approach in the current study.

3. Relevant research

For several years Norwegian authorities` have emphasized the significance of using ICT and digital learning materials in Norwegian compulsory school. Nevertheless, knowledge about the use of ICT and digital teaching materials is still very limited (Juul et al., 2010; Norwegian Research Council, 2008), and even more limited regarding the use of assistive ICT for disabled pupils (Murchland & Parkyn, 2010). Therefore, the following overview of relevant research will look into research about (i) the use of ICT and digital teaching materials, (ii) the significance of ICT for disabled children and young people, and (iii) the significance of assistive technologies for disabled children and young people.

3.1 Use of ICT and digital teaching materials in Norwegian schools

Even though ICT are widely disseminated throughout Norwegian society, and especially among young people, it turns out that Norwegian compulsory schools lag behind this development. Furthermore, a digital differentiation among pupils in compulsory school is documented (Arnseth et al., 2007; Hansen et al., 2009).

Digital differentiation denotes variables in the use of ICT among those who supposedly have equal access to ICT. More importantly is the focus of digital differentiation on the implications of these differences (Peter & Valkenburg, 2006; Sassi, 2005; Yu, 2006). While a digital differentiation perspective perceives the characteristics of the users as more important for his/her use of ICT than the characteristics of the ICT, studies on young disabled people finds the characteristics of the ICT to be of vital importance for their use of ICT (Söderström, 2009a, 2009b; Söderström & Ytterhus, 2010).

Available research shows that Norwegian compulsory schools employ ICT and digital teaching materials as an integrated part of the teaching only to a very small degree (Kløvstad, 2009; Juuhl et al., 2010; Vavik et al., 2010). While approximately 20 percent of teachers in compulsory schools use ICT regularly, the majority state that they use ICT "sometimes." The reason for this low use of ICT is because teachers lack experience with ICT, lack technical and educational support, and experience a shortage of time and equipment (Vavik et al., 2010). Moreover, there is very little research on how decisions are made when teachers choose which learning materials to use in their teaching (Juuhl et al., 2010). Available digital learning materials are only occasionally universal designed, and are inaccessible to disabled pupils, for the most part (Begnum, 2008; Bekken, 2009). These circumstances point to the need for more research on how the school may employ ICT, and especially assistive ICT, to implement the principle of inclusive education.

3.2 The significance of ICT for disabled children and young people

Western people, and especially young people, use objects, technologies and ICT equipment as symbols of identity and belonging. However, to display a desired self-identity, people have to use adequate symbols in appropriate ways, i.e. in ways that are culturally and historically contextualised (Buckingham, 2006; Hocking, 2000). In this way the use of technology in general and of ICT in particular, may promote or inhibit appearance, performance, and identity (Haraway, 1991; Latour, 1992; Söderström, 2009b). While this applies to most people in general, it is especially the case for children and young people.

Disabled children and young people are first and foremost ordinary children and young people with the same desires, aspirations, and needs as any other children and young people. The significance of ICT for disabled children and young people is quite similar to that of other children and young people. This means that ICT is above all valued for its social, interactive and communicative potential in networking and identity negotiations (Lonkila & Gladarev, 2008; Storsul et al., 2008; Söderström, 2009a, 2009b). However, these characteristics of ICT lead to a permeability of the virtual and the material world. This means that the material world and the virtual world are no longer separate entities, but are permeable, mutually constituted, and embedded in s young peoples` everyday lives (Buckingham, 2006; Peter & Valkenburg, 2006; Söderström, 2009a). This permeability involves some challenges in disabled children and young peoples` participation and inclusion in the peer group. One of these challenges is connected to the accessibility and usability of ICT. While accessible ICT is ICT a person can operate, usable ICT is ICT a person can use for his or her intended purposes (Söderström, 2009b).

Another challenge is the permeability of the virtual and the material, which leads to a dependency on usable ICT (McMillan & Morrison, 2006). This dependency is a double-edged sword because it embraces more and more of young people`s everyday lives. Consequently, the social, interactive and communicative use of ICT is vital for young people`s social inclusion and participation in the peer group. Furthermore, withdrawal from the use of ICT, or the inability to engage in digital forms of communication, is perceived to represent one of the most damaging forms of exclusion for young people (Livingstone & Helsper, 2007; McMillan & Morrison, 2006; Söderström, 2009a). Therefore, accessible and usable ICT holds great significance in disabled children and young people`s lives.

3.3 The significance of assistive technologies for disabled children and young people

Assistive technologies are technologies used to improve, expand or extend people`s performances, actions and interactions, and thus they are often experienced as an extension of the body (Lupton & Seymore, 2000; Moser, 2006; Winance, 2006). However, using assistive ICT involves more than overcoming environmental barriers; it also involves symbolic, historical and cultural contexts. Assistive technologies are loaded with collective cultural traditions, symbols and values, and subjective feelings and meanings are assigned to the technology (Pape et al., 2002; Wielandt et al., 2006). While young disabled people find the use of ICT to symbolise competence, belonging, and independence, they find the use of assistive ICT to symbolise restriction, difference and dependency. Therefore, to pass as ordinary young people, they reject using assistive ICT whenever possible. However, those disabled young people without the option to participate online without assistive ICT make a "forced choice" to adopt these technologies. Disabled young people who adopt assistive technology seek to make the technology coherent with their preferred self-identity, and if this is not achieved they reject the technology (Söderström & Ytterhus, 2010). This contradiction makes the combined use of ICT and assistive ICT of special interest to disability studies, the field of the current study.

4. Theoretical perspective

The current study draws on a perspective rooted in social science and the cross-disciplinary field of Nordic Disability Studies, and therefore it holds a relational perspective on disability. This perspective conceptualises disability as a social construction, which takes place in interpersonal relationships, in encounters between individuals and environments, and between individuals and society (Gustavsson et al., 2005). Consequently, this chapter's perspective is that it is not the disabled pupils' impairments as such that governs their use of ICT or assistive ICT, but rather the social practices in which the use of these technologies takes place.

Furthermore, this study employs an actor-network theory (ANT) perspective to elaborate on its findings. This perspective seeks to reveal what is happening, how it is happening, and what is involved in that which is happening (Moser, 2003). This perspective refuses to make *a priori* distinction between entities and actors, or define in advance what kind of entities might be granted agency and explanatory force. According to an ANT perspective one should have as few assumptions as possible about what there is in the world and how the different entities in the world are related. Also, one investigates what something is by asking what and how it is made to be, what its possibilities are and what relations it emerges from, what is done in practice, and what effects it brings along (ibid.). These conditions materialise in *socio-material practices* in which facts, objects, and nature are not given but are effects of interactions, relations and orderings. Such socio-material practices may be how disabled pupils use their assistive ICT, how the assistive ICT influences a disabled pupil's possibilities for actions and interactions, and how these possibilities affect the disabled pupil. Or in other words; how a disabled pupil's actions, participation, and identity are made and unmade in specific relations and interactions holding both social and material elements. This chapter investigates such socio-material practices in the Norwegian compulsory school.

The actor-network theory was originated by Latour (1987). He claimed that objects, machines, technologies and humans are all equal parts in reciprocal networks of connections and joining actions, all actively influencing each other, and all being actors (Latour, 2008). *Actors* generate effects. Therfore, any object, artefact, or person who generates an effect by making a difference is an actor. Actors may indicate, encourage, permit, influence, make possible, determine, or obstruct actions. Therefore, who and what enter into an action, or *social practice*, need to be carefully scrutinised. While human actions, communication and symbols only constitute one part of social practices, things, objects and technologies constitute the other part. In social practices the connections and joining of actions create network effects that constitute *social structure*. Follow the actor, reveal their actions, and show how the social is created says Latour (ibid.). What I find promising in an actor-network perspective is its sensibility to the ways in which technologies are made to be and people are made to do.

However, the ANT perspective does have its critics. Some have attacked it for being a reductionist perspective, while others are disturbed by its liberal argument for extending the notion of actors to non-human entities. Some critics accuse the perspective of being too preoccupied with productivity and network building, and consequently losing sight of contradictions, ambivalence and complexities. Feminist researchers in particular used in this criticism of an ANT perspective. However, studies employing the ANT perspective are executed in so many different locations and contexts that the perspective is constantly developing (Moser, 2003). Law (1994) is one of the present ANT perspective's proponents, emphasising the perspective's constructivist elements and pointing out that the researcher adopting this perspective also takes part in shaping reality, just as the socio-material practices the researcher studies do (ibid.). While the ANT perspective has developed from focusing on network building and the production of objects, to focusing on complex socio-material practices and enactments, to the collectives which make these practices possible, and to the actions and identities they enable (Moser, 2003). In this chapter I use this latter version of the ANT perspective to investigate how the use, or non-use, of assistive ICT in schools influence disabled pupils' opportunities for active participation in an inclusive educational system within the regular Norwegian school system.

5. Methodological approach

This chapter draws on a qualitative pilot study with three disabled pupils, their parents and teachers, and two employees at an assistive technology centre. The participating pupils are between 10 and 15 years of age, and they attend primary or secondary ordinary local schools in Norway. One of the pupils uses an electric wheelchair and has considerable mobility difficulties, one of the pupils is visually impaired, and the last pupil is blind.

While this study is a pilot study it was vital to recruit a purposeful sample. A purposeful sample selects potential information-rich cases for in-depth study and, therefore, maximises the potential for discovering as many dimensions and conditions related to the investigated phenomenon as possible (Patton, 2002; Strauss & Corbin, 1998). Thus, the current sample was selected after purposeful criteria about who might best generate theory. The participants were voluntarily and anonymously recruited through an assistive technology centre. The two groups of disabled pupils were chosen due to previous research that points out that people with mobility impairments encounter few ICT-related barriers, while visually impaired people encounter the most ICT related barriers (Fuglerud, 2006; DCR, 2004). Therefore, these two groups of youth might represent different aspects of knowledge to the study. All participants gave their informed consent, and they were assured full anonymity and confidentiality.

The data collection took place as individual semi-structured qualitative interviews with the three pupils, their parents, teachers and the employees at the assistive technology centre. Each interview lasted for approximately one hour. An interview guide was used to ensure that the same basic lines of inquiry were pursued with each participant (Patton, 2002). The subjects of the interviews were the participants` experiences of use of assistive ICT in the school, and what they thought might promote or hamper the use of assistive ICT in school. In addition to the interviews, participant observations were made of the three pupils` use or non-use of assistive ICT during a school day. Notes were taken during the interviews and the observations, and complementary notes were taken right after each interview or observation.

The analysis started right after the first interview was conducted and continued throughout the data collection period and the writing of this chapter. The data collection took place during autumn 2010. Data were analyzed using a *constructivist grounded theory* approach characterized by a continuously content comparative method. While traditional grounded theory is criticized for being too positivists and reductionist a *constructivist* grounded theory adopts grounded theory (GT) guidelines as tools but does not subscribe to the objectivist, positivist assumptions in GTs early formulations. The *constructivist GT* approach details processes and contexts and goes into the social world beyond the investigative story (Charmaz, 2005). These guidelines for analysing the data were very much in line with the theoretical perspective of the actor-network-perspective (Latour, 2008) employed in this study.

Even though this process is inductive, no qualitative method rest on pure induction. All knowledge arises through interpretation of the data: theoretical analyses are interpretive renderings of a reality and not solely objective reporting of it (Charamz, 2005). Out of this continuous content comparative interpretive approach the following categories emerged: (i) technological properties, (ii) interdisciplinary collaboration, and (iii) school administration and management. These categories will be expanded and discussed in the following section on the study's findings.

6. Findings

The findings that are described and discussed in this section are presented using case descriptions and quote excerpts. All descriptions are real and all excerpts and correct. The description and excerpts used are selected because they illustrate circumstances which are common for all participants. However, to ensure the participants` anonymity, some personally identifiable details are changed. For the same reason all pupils are referred to as *she*, the teachers are referred to as *he*, the parents are referred to as *mother*, and the employee from the assistive technology centre as *they*. All three pupils participating in this study were assigned a personal computer, both at school and at home, with the necessary assistive technologies to make the computer accessible for them. We will now look closer into how the participants experienced the use of assistive ICT in school and what they thought of their experiences.

6.1 Technological properties

Sometimes the technologies the disabled pupils were assigned worked just fine, and sometimes it did not work as they were supposed to work, and sometimes they did not work at all. The question is what the consequences of this are for the possibilities of inclusion of the disabled pupils and their participation with classmates.

6.1.1 Useful and compatible assistive ICT

One of the disabled pupils participating in the current study, Lisa, is a girl in fifth grade with severe mobility difficulties. Lisa uses a lot of assistive ICT at school. Sometimes she receives her teaching in class along with her classmates, and sometimes in a separate room alone with one teacher.

When Lisa is learning mathematics she attends her regular class. Lisa sits in an electrical wheelchair at a large desk in the front of the class. The teacher asks the class to solve the math tasks on some specific pages in the math book. A teacher`s assistant helps Lisa start the math program on her computer, and Lisa uses a joystick to navigate the marker on the screen to solve the same math tasks as the rest of the class. One by one the math tasks appear at the screen, proposing several possible answers. Lisa uses the Joystick to click on the answer she thinks is correct. Clicking on the right answer she is given points, more points for more difficult math tasks, clicking on the wrong answer she is given no points. Lisa navigates the marker quite quickly around on the screen using the Joystick. Because she has some involuntary movements in her upper limbs it is sometimes a little bit hard for her to stop the marker exactly at the correct answer. However, most of the time she stops the marker at the correct answer, and at the end of math class she has a lot of points. Lisa proudly shows all her point to some of her class mates, who stop at her desk and compliment her before storming out to recess.

In this math class the assistive ICT Lisa used functioned as an actor enabling her to be in control of things, participate in class, and to show a positively valued identity as a competent pupil. In this way the compatible software mathematics program and the usable Joystick functioned as an actor empowering Lisa in her education, and making her no different from the other pupils. Through the provision of this assistive ICT in this setting Lisa was made able to demonstrate her competence in "doing being ordinary" and thus, pass as ordinary. In as much as people do every day ordinary things people pass as ordinary (Goffman, 1963). These properties of the assistive ICT, the usefulness and the compatibility, are vital prerequisites for disabled pupils` possibilities to participation and inclusion in ordinary schools. When the assistive ICT works as expected the disabled pupils find them very intriguing. When asked about what they think of assistive ICT expressions such as *"I think ICT is an ingenious invention", "I would be lost without it"*, and *""It would be a boring life without it"* illuminates this technologies' central role in the disabled pupils` everyday lives.

However, action is not an activity done by one actor alone. Action involves many actors, human and non-human, in a network of connections (Latour, 2008). In that respect the possibilities of actions provided the disabled pupils by useful and compatible assistive ICT are made possible by a set of actors in network of connections. Lisa`s empowerment in math class is, thus, not solely due to the properties of the technologies used. It is also due to the combined interaction between, and properties of, Lisa, the assistive ICT, and the classroom setting. Thus, Lisa`s empowerment is made possible through three interrelating circumstances; (i) her mastering of the useful and compatible assistive ICT, (ii) her presence and participation in the classroom, and (ii) her classmates recognition of her competence.

I find this investigation of what is done in practice, what relations it emerges from, and what effects it brings along to illuminate how the *socio-material practices* in which the these things takes place are not given but effects of socio-material interactions, relations and orderings. Or, in other words what relations, people and objects interrelate and contribute to the full inclusion of, in this case Lisa, in the regular local school. This investigation of what actually takes place and what effects are brought along when Lisa employs useful and compatible technology is quite illustrative for all of the three disabled pupils in this study. However, it is not always that the assistive ICT is useful or compatible; quite frequently it turns out to be errors and shortcomings of the assistive ICT.

6.1.2 Errors and shortcomings of the assistive ICT

While Lisa sometimes experiences the assistive ICT as useful and compatible she just as often experiences it having errors or shortcomings. For writing and reading Lisa is assigned the assistive technologies MyTobii, which is an eye tracking software program for the computer, and BigKeys, which is an enlarged keyboard for the computer. Lisa enjoys writing and reading, she is expresses herself very well, and she wants to become an author. However, due to quite a bit of involuntary head and upper limb movements it is quite strenuous for her to focus her eyes long enough at one particular point when using the MyTobii, or to control her finger movements to hit the right key at the BigKeys. Consequently she needs a lot of time to solve reading and writing tasks which really is quite simple for her and it is quite exhausting for her to work on the computer over some time.

Lisa receives her writing and reading lessons in a separate room alone with one teacher. The assignment today is to write about the night before when she attended the local youth club. Lisa chooses to use MyTobii for this assignment, and she and her teacher sit close side by side at the computer. Lisa concentrates on focusing her eyes on the letters she wants to write in order to tell her story. It takes quite some time to write a whole sentence, and after a while she gets tired and they switch over to using the BigKeys. Every time Lisa tries to write the letter "A" nothing happens. She tries over and over again, but nothing happens. Lisa gets very upset by this, and after some time the teacher also gets a little bit frustrated. The teacher tries to figure out what is wrong, and after a while he discovers that the key for "A" really functions as a "delete" key. Both Lisa and the teacher appear to be happy to figure this out, and Lisa continues to write her story. But now she deliberately tries to avoid writing any word with the letter "A" in it. This is not easy, and it takes some creativity on Lisa's behalf. After approximately 45 minutes (one class hour) Lisa has written four or five sentences about the local youth club she went to the night before.

When the writing lesson is over the teacher asks Lisa if she wants to print out her story and share it with the rest of the class later that day. Lisa thinks this is a great idea, and they start to print out her story. However, it turns out that the printer which is supposed to be connected to Lisa`s computer is not connected after all. The teacher says this is the responsibility of the school`s IT manager, and that he himself is not able to do anything about this shortcoming. They have to notify the IT manager and wait for him to connect Lisa`s computer to the printer. Consequently, at the last class hour that day, when everyone is supposed to share their stories with one another, Lisa is not able to share her story with her classmates.

The missing letter "A" at the enlarged keyboard BigKeys was an obvious error of the assistive ICT, but whether the missing connection between the computer and the printer was an error or a shortcoming is hard to tell. Regardless of what the technical barrier consisted of it constituted a barrier in Lisa`s participation in her class. Thus, in this setting the technical error or shortcoming excluded Lisa from her classmates` sharing of their writings, and placed her outside the classmates` fellowship. Such consequences of technical errors and shortcomings are usually ascribed the disabled person, and not acknowledged as the consequence of the technology`s property (Söderström, 2009b). Lisa`s mother tell about the same errors or shortcomings of the same assistive ICT Lisa is allocated at home;

> *"When the letter "A" is not working this represent a barrier for Lisa. The missing "A" means that she has difficulties reading and writing e-mails and chatting with friends, thus she misses out on a lot of information and interaction. The assistive technology equipment just has to work. Why do they not check this out at the assistive technology centre before delivering it to us? Do they not expect her to need the letter "A", what do they think she uses the computer for? It is vital for her quality of life that all her assistive ICT just work all the time".*

Here Lisa`s mother describe how little strokes fell great oaks, and what consequences this may have. An assistive technology`s usability is reflected by its impact on the user's activity and social participation (Arthanat et al., 2007). To obtain a high level of usability assistive technology must reduce physical, cognitive, and linguistic efforts, promote convenience, efficiency, and productivity. And even more importantly, it must support a positive impression of the user on significant others.

Assistive technologies which do not work as anticipated or needed creates frustrations which very often lead to the rejection of the technologies (Pape et al., 2002; Ravneberg, 2010; Söderström & Ytterhus,, 2010). Especially children and young people have low tolerance for technical errors or shortcomings which hamper their self-presentation or interactions with peers. Encountering such technological barriers disabled children and young people feel that their impairment is placed on the front stage to use Goffmans terminology (Goffman, 1963), and this is something they strongly want to avoid. Consequently many of them avoid using any technology which makes them stand out as disabled, especially in their interaction with classmates and peers. As far as disabled children and young people have a choice most of them will chose to manage without using assistive ICT, because they experience the use of assistive ICT as stigmatising. However, many of these children and young people do not have this choice if they want to participate and be included in the peer group.

One of the other participants in the current study, let us call her Anna, is visually impaired, having a progressive condition. Some time ago Anna was assigned a lot of sight enhancement assistive technologies. She did not want to use them because she felt embarrassed using them, feeling they ascribed her as different from her classmates. After some years her condition became worse, and now she realises that she has to use assistive ICT to be able to do any schoolwork at all. Anna`s teacher comments that he thinks Anna has gone through a maturation process which has helped her realise the long term consequences of not using the assistive ICT. Adopting assistive technology is a time consuming process which very often is conflicting, and also associated with individual functioning and maturity level (Craddok, 2006; Söderström & Ytterhus, 2010). However, even when the assistive ICT is well integrated in the disabled pupils' everyday activities and all technical details are in place, there still occur barriers to overcome.

6.1.3 Useless and incompatible assistive ICT

Anna is now in seventh grade, and she very often experiences her assistive ICT to be incompatible with ordinary ICT. At school Anna uses the following assistive ICTs screen magnifier, screen reader, speech synthesis, and Braille display. She also uses assistive technologies such as whiteboard camera and reading TV at school. Anna enjoys school and takes pride in her homework. She attends her regular class the whole time at school, but because she needs a lot of space for all her assistive technologies she is seated a bit on the side of the rest of the class. Her classmates is seated in pairs at one desk, but Anna needs two desks by herself in order to make room for all her technologies. When I ask Anna what she thinks about using all these technologies at school she answers:

> *"I don`t mind. It is convenient to have all these assistive technologies, but sometimes it is quite tiring too. Especially with the screen magnifier and the screen reader… For the most part I really use the school`s desktop PC".*

Why Anna for the most part uses the school`s desktop PC when she is allocated her own portable PC is illustrated by the following case excerpts.

At a history class the pupils are asked to choose one of the history's famous explorers, to gather information about the person, and then write an essay about this person. For this assignment they have two history classes (90 minutes). Anna`s group goes to the computer lab. Each pupil sits down with a desktop computer and starts surfing the web, looking for information about the one explorer they want to write about. Even though Anna is allocated her own portable PC and the assistive technologies necessary to make the PC accessible to her, she now chooses to use the schools ordinary desktop PC in the computer lab. The teacher explains that the assistive technology centre quite recently have installed a new software program called Supernova on Anna`s portable computer. Unfortunately it turns out that after this new software program has been installed Anna`s computer is no longer compatible with the web.

Nevertheless, Anna embarks with great enthusiasm for the task she is given. She decides to write about the Norwegian explorer Nansen, and starts searching the web for information about him. However, to be able to read anything on the school`s computer screen she has to enlarge the font to the size of font 36. When she tries to read the enlarged text on the screen she sits with her nose almost inside the screen and reads one word at the time. Reading on the screen Anna has to move along one line at the time, and word for word. This takes a lot of time, and many times she has to go back and read the line once more, probably to get the context right. After a long time spent on reading Anna writes one sentence about Nansen fairly quickly, even though she has to stare hard on the keyboard to hit the right keys when she writes. After writing one sentence she goes back to searching the web for more information.

The menu keys at the top of the screen can, however, not be enlarged, and Anna is not able to find the right menu key when she needs one of them. One of Anna`s class mates is sitting right next to her at the computer lab, and every time Anna needs one of the menu keys her class mate helps her out by point directly at the menu key Anna needs. During the two history classes (90 minutes) spent at the computer lab Anna manage to write seven sentences about Nansen. The most of this time Anna has spent on searching the web, moving along one line on the screen, reading one word at the time about Nansen. When her class mates tells Anna they have to stop because the time is out Anna gets very stressed saying; *"But I am far from done"*. Her class mate, who have finished her assignment a long time ago, says; *"Well done, you are clever"*.

This case excerpt illustrates how incompatible technology creates barrier in disabled pupils school work. Furthermore, incompatible technology creates barrier for Anna`s possibilities to show her competence and similarity to class mates. During the two history classes Anna was only made able to write seven sentences about Nansen. This was not due to her having poor writing or reading skills, but to inaccessible menu keys on the computer and almost unreadable web sites. If her allocated portable PC had been compatible with the web Anna could have acquired information about Nansen much more quickly, solved her task in time, and proved herself as a competent seventh grader. Anna`s class mate`s comment at the end of the session about Anna being clever might suggest that the class mate do not expect Anna to produce the same amount of text as she herself and the other class mates did. Or, maybe she did understand what a struggle it was for Anna to work on an inaccessible computer?

The relation between universal design of ordinary ICT and individual adaptation of assistive ICT is central in research on disabled people and ICT. While some researcher focus on the inaccessibility of ordinary ICT to promote the need for flexibility and universal design, other researchers argue for the necessity of individual adaptations of assistive technology, and others calls for the combination of the two. However, it turns out that such experiences with incompatible technology creating barriers for activity, just as Anna experienced in history class, are quite common (cf DCR, 2004; Söderström, 2009b; Söderström & Ytterhus 2010; Wielandt et al., 2006). For assistive ICT allocated disabled pupils to promote enhanced inclusion and participation in school settings these technologies have to be compatible with the ordinary ICT used at school. Moreover, for this to work satisfactorily it requires commitment from several sources, and close interdisciplinary cooperation.

6.2 Interdisciplinary collaboration

To ensure disabled pupils the benefits of using ICT tools in school life requires a concerted effort from many different actors. Regular testing, customization, training, operation and maintenance of equipment are prerequisites for appropriate usage of the equipment. Actors involved in this work are local occupational therapists, assistive technology centre, various centers of expertise, technical suppliers, teachers, the local school`s IT manager, and the municipality. To coordinate efforts of all these actors through interdisciplinary collaboration proves to be difficult. In this pilot study especially two aspects of collaboration around the use of the individual pupil's ICT tools turns out to create barriers to pupil participation and inclusion. The first thing we will investigate is the consequences of things taking time.

6.2.1 Things takes time

The third pupil who participated in this pilot study is a severely visually impaired girl in ninth grade. Let us name this girl Eve. Eve is a clever pupil and she is allocated the same assistive technologies as Anna is. As the other disabled pupils Eve has a double set of assistive ICT, one set at school and one set at home, and she does virtually all school work using the assistive equipment allocated her from the assistive technology centre. As long as all the equipment works all goes well, but as soon as something doesn't work everything falls apart. Eve is totally dependent on her ICT tools in order to do school work. When an error occurs on the assistive ICT the assistive technology centre is responsible of repairing this (Svendsen, 2010). Eve`s mother tells:

> *"It feels like it takes years from the time we send in assistive equipment to the assistive technology centre until it is repaired. In the meantime, Eve must do without this equipment and this has major consequences for her schoolwork. I get so tired of calling in to the assistive technology centre and nag at them, it's so tiring! And then it's a little hard to complain too, for we are really grateful for all the help we get. But it is so much that is missing, and we're totally dependent on them."*

That things take time is something most of us have experienced, especially where multiple agencies have to collaborate on complex matters. This also applies to the dissemination and maintenance of assistive ICT equipment. This is confirmed by many users of this sort of equipment, and also by various service providers in the municipalities. Especially is information about what happens along the way, where the equipment is, and how long it takes before it is repaired something which is missing (Svendsen, 2010). Many disabled pupils are completely dependent on their ICT tools to work at any time in order to do their school work. When the parents have to take responsibility and to make calls to check and nag, it feels both frustrating and tiring for them. They also experience it as a dilemma to complain about a service they basically are not happy with, but which they are totally dependent on, and also thankful for. The assistive technology centre confirms that it takes a long time to repair assistive ICT, and they explain:

> *"When the assistive technology is sent to be repaired it is being sent directly to the engineers at the assistive technology centre, without notifying us who provide the users with the necessary equipment. We want to be notified to follow up the user when he or she is without any assistive ICT, or to provide them a replacement for the equipment while they are waiting. These things we have discussed here at the assistive technology center, we want to do something about it, but so far nothing has happened".*

However, it is not only the assistive technology centre which is involved in maintaining the assistive ICT, and even though it takes a long time to get the equipment repaired just to find out who is supposed to repair the equipment may take just as long time. One of the teachers explains:

> *"It's been a lot of back and forth with the assistive equipment for this pupil, not least when it comes to figuring out what to do when the equipment does not work. For instance; it is the IT manager at the local municipality who is responsible for providing access to the web. The printer, which, for the time being does not work, is the responsibility of the IT manager at our local school, and the eye tracking program is it the technical supplier which is responsible for, while the enlarged keyboard is the responsibility of the assistive technology centre. Whatever can be fixed here at our school runs more smoothly, but anything that needs to be done by the municipality, the supplier or the assistive technology centre takes an incredibly long time."*

The teachers in this pilot study all report that they spend a lot of time figuring out things which they feel it really should be clear procedures for, such as who is responsible for what. This leads not only to the fact that many things take longer time than necessary, but also that many other important things are not done.

6.2.2 Unclear responsibilities

One of the teachers wants to show how the braille display on Anna`s laptop is working. After much trial and error he must give up this attempt. It seems like he is a little confused and unsure whether it really is the Braille display that is not working properly, or whether it is actually he himself who do not remember quite hoe to use it. Somewhat resigned he exclaims:

> *"The hardest part is actually finding out who is responsible for what when it comes to maintenance, upgrading and training in using the equipment. That is how the formal division of responsibility is between the school, the assistive technology centre and the resource centre".*

This teacher started to work with Anna about one year ago. When he started to work with Anna he got no training or guidance in how to use her assistive technologies, and he still have not received any training in use of this technology. He is, however, a young and dedicated teacher and not afraid to try out new things, so he has taught himself how to use the most of Anna`s assistive technologies. In the beginning he felt it a little overwhelming as it was so much equipment to being so much equipment to learn to use at once. It took really a long time before he felt he had some kind of overview. To be able to do his job reasonable proper he believe it has been imperative that he is not afraid to try out new things, and that he is interested in technology. Both the assistive technology centre and resource centre offers various courses in use of the assistive technology equipment that Anna uses. However, the teacher has not received any information about these courses, and he is not aware that this service exists. It is one of the other teachers who are responsible for providing training and education for teachers when needed. This teacher is also responsible for providing information about relevant courses. This information has, however, for some reason Anna`s teacher not received. Usually it is not the lack of relevant courses which is the problem, but rather that the teachers don't get the opportunity to attend them (Bekken, 2009). According to this pilot study this might be due to lack of information about relevant courses.

The assistive technology centre organizes a lot of courses about use of various assistive ICT. These courses are intended for teachers, parents and other persons working close to the pupils who use assistive ICT. The courses are free of charge, and the assistive technology centre distribute information about their courses to the local municipality`s contact person for assistive technology. This contact person is then supposed to disseminate this information to teachers, parents and other persons working close to the persons using assistive ICT. This means that it is of vital importance that the contact person for assistive technology in each municipality is able to keep track of everyone who is assigned assistive technologies. It appears that this is not always the case. In addition information about the courses has to go through several segments before reaching its target. In our case first from the assistive technology centre, through the local contact person, then to the local school, and finally to Anna`s teacher. It appears that somewhere along this road the information get lost. Moreover, in cases where this information do reach the local school it is entirely up to the school if they want to make use of the courses they are offered or not. Even though the courses are free of charge, nevertheless, to attend these courses turns out to be an economic issue. If one teacher attends a course for one day the school has to hire a substitute, and this involve an expense. In times where budgets are thigh the schools have to make priorities, and courses and education are usually not prioritized areas when resources are scarce.

The assistive technology centre does not have the capacity to follow up everyone who is assigned assistive ICT. Thus, the courses they offer are the only arena for guidance and training in use of the assistive ICT equipment, which can be very complicated and specialized. The courses are also intended to provide an arena for sharing experiences, successes and frustrations. Thus, from the assistive technology center's point of view it is up to the individual person working close to the user of the assistive ICT how well they make use of the potential of the assistive ICT. However, this argument makes the presumption that everyone who needs it is given the opportunity to attend their courses. This is, however, not always the case, which the excerpts from Anna`s teacher on the previous page illustrates.

From the assistive technology center's point of view the cooperation with the teachers varies a lot. While they experience some teachers as very interested and eager to learn, others are quite reluctant and appear to have mental blocks when it comes to technology. Similar experiences are also known from other studies in the field, describing professional helpers either as enthusiasts who makes everything possible, or as plugs that stops everything (Bekken, 2009; Egilson, 2010, Lundeby & Tøssebro, 2006). Further, the assistive technology centre experiences the cooperation with the different teachers as dependent on the teachers working culture. That is what kind of subject or discipline the teacher belong to, or the workplace milieu the teacher is socialized into. Last, but not least, the assistive technology centre also finds the frames and structures under which the teachers work to be very influential on how the teachers are made able (or unable) to employ the potential of assistive ICT.

Unclear responsibilities are not only an issue for the cooperation between different departments. Clear distribution of responsibilities inwards at the individual local school is also a challenge. Such unclear distribution of responsibilities involve an extra strain for parents of disabled pupils in that they take on responsibility which actually is the school`s responsibility. One of the mothers expresses this as follows:

> *"In general the teachers lack creativity when it comes to adaptation of the curriculum to my child. It appears that they do not plan anything ahead. They tell me that they want to make the adaptation that is necessary, but that they have no time for this kind of work. Especially in math classes it is important that they make the necessary adaptations for her. It is so hard for her to figure out a calculation if she has to navigate herself through a divided text".*

Further on this mother tells that she constantly has to remind the teachers about what her child needs. This mother finds that all the teaching is organised for sighted pupils, and that the special teacher then has to help her daughter to acquirer what she can of this teaching. The mother thinks it should be an easy task to adapt the teaching for the class so that her daughter too could participate in math class together with her classmates. *"It ought to be easy to use enlarged text on the blackboard, sheets handed out, and power point presentations"* she says. This mother experience that she has to do half the job the school is supposed to do by her herself making the teaching material accessible to her visually impaired daughter. This is in accordance with Bekken (2009) who found that parents of disabled pupils put a lot of voluntary work into adaptation of curriculum, teaching material and technology. Unclear distribution of responsibility inwards at the individual local school does not only hamper cooperation with external partners, but it also involves an enhanced strain on parents of disabled pupils (Lundeby, 2008; Wendelborg, 2010). This is actually a question of the individual school`s incorporated routines about how to meet pupil`s individual needs. How the individual school is organised and administered is of vital importance when it comes to evaluating to what extent the school is an inclusive school or not (Wendelborg, 2010; Wendelborg & Kvello, 2010; Wendelborg & Tøssebro, 2010).

6.3 School administration and management

According to Tronsmo (2010) school administration is to take responsibility for achieving good results. Additionally, the school administration has an indirect impact on the teachers' level of ambition, the educational setting, norms and culture, in addition to the schools collaborative relationships with external partners (ibid.). How the school day is organised for pupils with disabilities has proven to be of vital importance for their possibilities of inclusion and participation with classmates (Bekken, 2009; Wendelborg, 2010).

At one of the schools in the current pilot study it appears that the way in which the school administration organizes the school provides enhanced possibilities for disabled pupils` participation and inclusion. There are several disabled pupils at this school, and the school administration is very keen on including all pupils regular teaching settings. *"The administration is very good at facilitate individual adaptation"* the teacher at this school explains. When the partially sighted pupil attended this school the administration told the teacher to let them know if he needed anything special in order to adapt his teaching to this particular pupil. In other words, the administration was not only aware of the pupils' potential special needs it was also very forthcoming and benevolent in its attitude to make an effort to include this pupil in the ordinary school setting. The teacher at this school has also been given the opportunity to get an education as a special educator for visually impaired pupils. Consequently this visually impaired pupil receives all her teaching in the classroom together with her classmates, which is a prerequisite for being included.

This particular school has gathered a number of disabled pupils in the same class. The teacher tells how all the teachers teaching in this class have been carefully chosen according to their ability to be flexible, creative and innovative. Nevertheless, even though the school administration appears to be very forthcoming, and the teachers highly qualified, the teacher participating in the pilot study experience that the other teachers lean a lot on him. He feels he has to take responsibility for many things that are not really his responsibility, e. g. universal design of classrooms etc. He thinks this is partly because he has some special education in the area, and partly because this is no one else's responsibility in particular. This illustrates the necessity of having a clear distribution of responsibility, inward in the individual school, and for the special areas of universal design and of individual adaptation. In total, it still appears that the school administration`s deliberate and positive focus on the issues of individual adaptation and full inclusion of disabled pupils is a prerequisite for the participation and full inclusion of disabled pupils.

At one of the other schools in the pilot study, however, it appears that the school organization and administration may hamper the disabled pupils` possibilities for participation and inclusion. The teacher at this school tells that there are no other disabled pupils at this school besides the one participating in this study. Universal design and individual adaptation for pupils with special needs are not an issue among the teachers. *"The school administration does not have a clue about disabled pupils' rights"* says this teacher, referring to the Education Act of 1997, which proclaim that all children have the right to attend their local school, and a regular class or group at this local school (Education Act, 1997). Even though the participant having mobility difficulties attend this school, which is her regular local school, it turns out that she rarely attends her ordinary class. Most of her teaching she receives in a separate room and alone with one teacher.

The argument for this exclusion from the ordinary classroom setting is that the assistive technologies this pupil uses requires a lot of space, and is best kept in a separate room. This argument is in line with the developmental trend of exclusion ordinary local schools which Wendelborg (2010) discovered. Wendelborg (2010) found that about 70 % of disabled pupils attending ordinary schools receive their education in separate settings, excluded from their classmates, in more than 40 % of the time spent at school. Moreover, Bekken (2009) found that when it comes to the exclusion of blind and partially sighted pupils this is due to the schools lacking competence, and the lack of follow-up at all levels. This development trend is quite contrary to the inclusive ideology of the Norwegian Education Act (1997) and other public documents in the area. In addition, this trend illustrates that no matter what potential assistive ICT holds for disabled pupils the technology itself do not contribute to enhanced participation and inclusion in the ordinary school setting. The triggering factor for the full utilization of assistive ICT`s anticipated benefits is first released when it is employed in an inclusive setting with classmates.

To be familiar with assistive ICT takes time and to learn how to utilize all its potentials takes time. Time for this kind of testing and preparation is, however, something the teachers feel they do not have. Even though some of the teaching is done by means of quite advanced technical equipment none of the teachers in this pilot study are given any extra time to prepare how to use this advanced technical equipment. This means that very often the pupil`s assistive ICT is not fully utilized, and sometimes not used at all. The non-use of available technical equipment is also well known from research on teachers' use of technology in general (Vavik, et al., 2010). It appears as if the introduction of new technological equipment provokes insecurity and that many teachers have an indefinable resistance towards employing new technology. The parents in this study all strongly point out that they experience the teachers' technological competence as insufficient. One of the younger teachers tells, however, that he has to use some of the time intended for teaching to get acquainted with the assistive ICT. This is absolutely necessary he claims, in order to be able to use it, but this time is also valuable teaching time which is lost. The question is if the school administration`s organization of the teachers` time, the teaching and the pupils` schooldays is a question of resources, of competence or of attitudes?

When a pupil is allocated a new kind of assistive ICT the pupil`s teacher is offered an introduction in how to use its core functions, either by the assistive technology centre or by the supplier of the technology. However, in order to be able to make use of all its possibilities the teachers say they need regularly courses and guidance in use of the technology. Such courses are offered twice a year by the assistive technology centre, though without any of the teachers in this study being aware of this. To disseminate information about these courses throughout all levels of services is an organizational challenge, a challenge which obviously is not met. The school administration`s lack of focus on individual adjustment and the significance of advanced assistive ICT have an impact, not only on the teachers premises and competence in this area, but also on the disabled pupils' possibilities for participation and inclusion in school. It might be that school administrations perceive this as a resource question, or as a question of priorities. It most certainly is a question of organization and management. Even though some teachers do a very good job in adjusting the teaching to the individual needs of their pupils they are still quite dependent on the school administration`s support. Securing individual adjusted teaching to disabled pupils requires a combination of thorough planning, collaboration, competence and the full utilization of individually assigned assistive technologies.

7. Perceived barriers and recommended solutions

All parents participating in this study accentuated the lack of competence in school when it came to assistive technologies. Further they pointed to lack of focus on the significance of individual adjustment of the teaching material and teaching situation. The parents pointed to these two conditions as the two greatest barriers for the full employment of the potential in assistive ICT. Even though the parents experience that some teachers do their best, the parents think this is still not enough. The parents think it is important that the teachers can be good examples to their children, also when it comes to using technology. This includes making use of assistive ICT an ordinary task in line with use of any other kind of technology. *"Actually, school administrations` competence regarding individual adjustments and inclusion of disabled pupils are in general quite inadequate and especially inadequate when it comes to the significance of assistive technologies"*. This is a statement re-occurring in the interviews with the parents.

The teachers also experience encountering barriers in their work with pupils who are allocated assistive ICT. They accentuate the lack of time to prepare and learn how to use the assistive ICT as the greatest barrier. Moreover, the teachers point to technical properties as barriers, such as non-working keys and missing software programs. One of the teachers also point out that resistance among teachers to employ new and unknown technologies also represent a barrier.

From the assistive technology center`s point of view it is the lack of environmental accessibility that represent the grates barrier in using assistive technology in school. They think the lack of universal design of school buildings and classrooms hamper the full utilization of the potential of assistive ICT, and of inclusion of disabled pupils in classroom settings. They also turn the focus in another direction; the inclusion of non-disabled pupils in the teaching of disabled pupils. This kind of a somewhat reverse thinking might also facilitate the inclusion of disabled pupils according to their point of view. Eventually they also accentuate the significance of teachers' competence, attitudes and experiences in the area as a possible barrier.

Parents, teachers and the assistive technology centre all experience barriers in using assistive ICT. They experience these barriers somewhat different, and they perceive the causes of these barriers different. However, all of them express the same concern; the lack of technological and digital competence in using assistive ICT. The parents emphasize the significance of school administrations` attitudes towards disabled pupils, the teachers emphasize the lack of time for preparations, and the assistive technology centre emphasize the practical adjustment and inclusion of non-disabled pupils in the teaching of disabled pupils.

The recommended solutions also differ according to ones point of departure. The parents think the solution must be providing all teachers enhanced competence in use of technology and in individual adjustment of the teaching. The teachers think the solution must be providing them more time for preparation, planning and testing of the assistive ICT. And the assistive technology centre thinks the solution must be a much more active effort in adjusting the environments to include disabled pupils in the ordinary classroom setting.

While the findings in this study to a certain degree is colored by the participants' different relationships to the participating pupils, the finding also correspond with previously research in this area (cf. Bekken, 2009; Craddok, 2006; Murchland & Parkyn, 2010; Söderström & Ytterhus, 2010; Wendelborg, 2010).

8. Conclusion

The subject of the current pilot study is disabled pupils` use of assistive ICT in Norwegian primary and secondary schools. Taking this pilot study as a point of departure this chapter has investigated how use, and non-use, of assistive ICT at school influence disabled pupils' opportunities for active participation in an inclusive educational system within the regular Norwegian school system. The findings elaborated on in this chapter illuminate the existence of barriers on several levels for the utilization of assistive ICT in Norwegian schools. Some of these barriers are technical barriers connected to the equipment, some barriers are human barriers connected to competence and attitudes, and some barriers are at a system-level connected to interdisciplinary collaboration and school administration.

The first barrier described is the technical barrier which occurs when the assigned assistive technology does not work as intended. This might be due to simple technical errors, or incompatible software or hardware. Such technical barriers put an effective end to any attempt of participation or inclusion of the disabled pupils in the teaching or the interaction between classmates.

Another barrier is the unclear distribution of responsibility among collaborative professionals, both between different levels of services and within the same service level. It is a huge amount of service providers involved in the provision of assistive ICT to pupils, e.g. the local occupational therapist, the assistive technology centre, the technical supplier, various resource centers, the local municipality, the local school, the teacher etc. While all these actors spend time figuring out who is responsible for what the disabled pupils who are dependent on assistive ICT is left at a dead end, so to speak.

This study illuminates how the individual school is organised and administrated is of great significance for disabled pupils' possibilities in participation and inclusion at school. School administrations` attitude towards making an active effort to include disabled pupils in the ordinary school settings set the framework for teachers` possibilities to make individual adjustments in their teaching. While the administration at one of the participating schools in the current study had a positive and active approach in including disabled pupils in the ordinary school setting, this was not an issue at all in the two other participating school administrations.

This study highlights phenomena and practices which have an impact on the use and non-use of assistive ICT in Norwegian primary and secondary schools. In sum these phenomena and practices turn out to be about teachers`, school administrations` and collaborative partners` knowledge, competence and attitudes towards technology and disability. These findings are also in line with previous research in this area (cf. Bekken, 2009; Craddok, 2006; Murchland & Parkyn, 2010; Söderström & Ytterhus, 2010; Wendelborg, 2010).

This current study points out how practical details and situations make a difference in disabled pupils` every day at school. When the assistive ICT works as intended and use of this technology is integrated in the teaching in ordinary classroom settings use of assistive ICT facilitate enhanced participation and inclusion of disabled pupils. It appears, however, that this integration of assistive ICT in ordinary classroom settings takes place only to a limited extent. When the assistive ICT does not work as intended, is not put to use, or is used only in separate and excluded teaching settings it is perceived as a symbol of difference which hamper participation and inclusion.

Further, there are two factors which emerge in this study which are found to be of special importance to underline and repeat in conclusion. First and foremost it is the significance of integrating use of assistive ICT for disabled pupils in the ordinary classroom setting. This integration will restrain potential stigmatization and subsequent rejection of the assistive ICT (Lupton & Seymour, 2000; Murcland & Parkyn, 2010; Söderström & Ytterhus, 2010). The other factor which is found important to underline and repeat is the assistive technology center's suggestion about turning the focus on inclusion into another direction; namely the inclusion of non-disabled pupils in the disabled pupils` use of assistive ICT. This suggestion is found very interesting, and might be a new strategy in promoting appreciation and inclusion of disabled pupils.

When it comes to what new knowledge this study provides the current pilot study highlights two conditions not described in previously research. First and foremost the study provides new insight into the consequences for the individual disabled pupil of the complexities of actors involved in allocating, operating and maintaining the assistive ICT. The second new insight this study provides is the consequences for the individual disabled pupil of her/his teacher`s perceptions of technology and disability.

The current study is a small pilot study and, thus, is perceived as a preliminary study in a relatively new combination of research area. The study`s intention is, however, to provide a basis and direction for further and more comprehensive research in this field. Besides investigating the current subjects on a larger scale, further research in this field should also look closer at the perspectives of school administrations, municipal authorities and the various service providers on technology and disability. In a Norwegian setting there still exist some gaps in the knowledge of what significance use of assistive technologies holds for disabled pupils in ordinary schools. This knowledge gap is especially present when it comes to how purposeful and efficient use of ICT and assistive technologies may provide new inputs to the principles of adapted learning environments and inclusion of disabled pupils. In this context it is important to note that how use of assistive ICT for disabled pupils creates possibilities, or barriers, in their participation and inclusion can only be revealed by closely investigating the cultural context and interactive processes in which the persons, technologies and environment are embedded. In practice this means to ethnographically study the phenomenon over time.

9. References

Arnseth, H. C., Kløvstad, V., Ottestad, G., Hatlevik, O. og Kristiansen, T. (2007) *ITU Monitor 2007 – Skolens digitale tilstand (ITU Monitor 2007 – The school`s digital condition)* Oslo: Universitetsforlaget

Bailey, J. (1998) *Inclusion through categorisation* pp. 171-185 in R. Booth & M. Ainscow (eds.), *From them to us: An International Study of Inclusion in Education* London: Routledge

Begnum, M. N. (2008) *Interaktivt læremateriell er i liten grad universelt utformet. Sluttrapport UTIN-prosjektet (Digital teaching material is rarely universal designed)* Oslo: MediaLT

Bekken, W. (2009) *Foreldres erfaringer fra samarbeid og tilrettelegging for barn med synshemminger i skole En kvalitativ undersøkelse,(parents experiences with collaboration for visually impaired children)* Oslo: Assistanse

Buckingham, D. (2006) "Children and New Media" pp. 75 - 91 in Lievrouw, L. A. & Livingstone, S. (eds.) *The Handbook of New Media Updated Student Edition,* London: SAGE

Charmaz, K. (2005) "Grounded Theory in the 21st Century: Applications for Advancing Social Justice Studies" in Denzin, N. K. & Lincoln Y. S. (eds.) *The Sage Handbook of Qualitative Research* Third Edition, pp. 507-535 London: Sage

Chiang, M. F., Cole, R. G., Gupta, S., Kaiser, G. E., & Starren, J. B. (2005) "Computer and World Wide Web Accessibility by Visually Disabled Patients: Problems and Solutions" *Survey of Ophthalmology* 50 (4): 394-05

Craddock, G. (2006) "The AT continuum in education: Novice to power user" *Disability and Rehabilitation: Assistive Technology* 1(1-2): 17-27

DCR (2004) *The Web Access and Inclusion for Disabled People A formal investigation conducted by the Disability Rights Commission* London: TSO

Education Act (1997) *Om lov om grunnskolen og den videregående opplæringa (About act on the compulsory school and the high school)* Oslo: Ministry of education and research

Egilson, S. T. (2010) "Parent perspectives of therapy services for their children with physical disabilities" *Scandinavian Journal of Caring Sciences* doi:10.1111/j.1471-6712.2010.00823.x

Fossestøl, K. (2007) "A policy for social integration. ICT strategies, employment and disability in the United Kingdom, the Netherlands, Denmark and Norway" pp. 1-10 in Fossestøl, K. (ed.) *Stairway to heaven? ICT-policy, disability and employment in Denmark, the Netherlands, United Kingdom and Norway* Oslo, NO: WRI-report 2007:5

Fuglerud, K. S. (2006) *Full participation for all? Developmental trends 2001 - 2006* pp. 91-140, Oslo, NO: Nasjonalt DOK og Sosial- og helsedirektoratet

Gustavsson, A., Tøssebro, J. og Traustadòttir, R. (2005) `Introduction: approaches and perspectives in Nordic disability research` i Gustavsson, A., Sandvin, J., Traustadòttir, R. and Tøssebro, J: (red.) *Resistance, reflection and Change Nordic Disability Research,* s. 23-44, Lund: Studentlitteratur

Hansen, L. S. (2007) "ICT policy in Norway - disability and working life" pp. 27-39 in Fossestøl, K. (ed.) *Stairway to heaven? ICT-policy, disability and employment in Denmark, the Netherlands, United Kingdom and Norway* Oslo, NO: WRI-report 2007:5

Hansen, I. L. S., Hernes, G., Hippe, J. M., Kalhagen, K., O., Nafstad, O., Røtnes, R. & Seip, Å., A. (2009) *Det norske IKT-samfunnet – scenarier mot 2025 (The Norwegian ICT-society – scenes towards 2025)* Sluttrapport i prosjektet IKT og samfunnsutvikling. Et fellesprosjekt mellom Econ Pöyry og Fafo, Fafo-rapport 2009:08

Haraway, D. J. (1991) "A cyborg manifesto: Science, technology and socialist-feminism in the late twentieth century" pp. 183 - 202 in Haraway, D. J. *Simians, Cyborgs and Women The Reinvention of Nature* London: Free Association Books

Hocking, C. (2000) "Having and Using Objects in the Western World" *Journal of Occupational Science* 7(3):148-157.

Juuhl, G. K., Hontvedt, M. & Skjelbred, D. (2010) *Læremiddelforskning etter LK06 Eit kunnskapsoversyn (teaching material research after LK06 an overview of the knowledge)* Rapport 1/2010 Høgskolen i Vestfold

Kent, B. & Smith, S. (2006) "They Only See It When the Sun Shines in My Ears: Exploring Perceptions of Adolescent Hearing Aid Users" *Journal of Deaf Studies and Deaf Education* 11(4): 461-476.

Kløvstad, V. (2009) *ITU Monitor 2009 Skolens digitale tilstand (ITU Monitor 2009 The digital condition of the school)* Oslo: Forsknings- og kompetansenettverk for IT i utdanning

Krumsvik, R. J. (2007) *Skulen og den digital læringsrevolusjonen (The school and the digital educational revolution)* Oslo, Universitetsforlaget

Latour, B. (1987) *Science in action* Cambrigde: Harvard University Press

Latour, B. (1992) "Where Are the Missing Masses? The Sociology of a Few Mundane Artifacts" pp. 225-258 in Bijker, W. & Law, J. (eds.) *Shaping Technology/Building Society* London: The MIT Press.

Latour, B. (2008) *En ny so cio lo gi for et nyt samfund Introduktion til Aktør-Netværk-teori (a new sociology for a new society Introduction to Actor-network.theory)* København: Akademisk Forlag

Law, J. (1994) *Organizing modernity* Oxford: Blackwell

Livingstone, S. & Helsper, E. (2007) `Gradations in digital inclusion: children, young people and the digital divide` *New Media & Society* 9(4):671-696.

Lonkila, M. & Gladarev, B. (2008) "Social networks and cellphone use in Russia: local consequences of global communication technology" *New Media & Society* 10 (2):273-293.

Lundeby, H. & Tøssebro, J. (2006) "Det er jo milepæler hele tiden" Om familien og "det offentlige" (There are milestones all the time) pp. 244 - 279 in Tøssebro, J. & Ytterhus, B. (eds.) *Funksjonshemmete barn i skole og familie Inkluderingsideal og hverdagspraksis* Oslo: Gyldendal Akademisk

Lupton, D. & Seymour, W. (2000) "Technology, selfhood and physical disability" *Sociel Science & Medicine,* 50(12): 1851-1862.

McMillan, S. J. and Morrison, M. (2006) "Coming of age with the internet: A qualitative exploration of how the internet has become an integral part of young people's lives" *New Media & Society* 8(1):73-95.

MD (2005) *e-Norge 2009- det digitale spranget (e-Norway 2009- the digital leap)* Oslo: Ministry of Government Administration, Reform and Church affairs

Moser, I. B. (2003) *Road Traffic Accidents: The ordering of Subjects, Bodies and Disability* Oslo: Unipub AS

Moser, I. (2006) "Sociotechnical Practices and Difference On the Interference between Disability, Gender and Class." *Science, Technology & Human Values* 31(5): 1-28.

Murchland, S. & Parkyn, H. (2010) "Using assistive technology for schoolwork: the experience of children with physical disabilites" *Disability and Rehabilitation: Assistive Technology* 5(6): 438-447

Norwegian Research Counsil (2008) *Programplan for utdanning 2020 Norsk utdanningsforskning fram mot 2020 (2009 - 2018) (program for education 2020 Norwegian educational research towards 2020*

NOU 2001:22 (2001) *Fra bruker til borger En strategi for nedbygging av funksjonshemmende barrierer (From user to citizen A strategy for diminishing diabling barriers)* Oslo: Statens forvaltningstjeneste

Pape, T. L. B., Kim, J. & Weiner, B. (2002) "The shaping of individual meanings assigned to assistive technology: a review of personal factors" *Disability & Rehabilitation* 24(1/2/3): 5-20.

Patton, M. Q. (2002) *Qualitative Research & Evaluation Methods* 3.ed. London: Sage

Peter, J. & Valkenburg, P. M. (2006) `Research Note: Individual Differences in Pereptions of Internet Communication` *European Journal of Communication* 21(2): 213-226.

Ravneberg, B. (2009) "Identity Politics by Design - Users, Markets and the Public Service Provision for Assistive Technology in Norway" *Scandinavian Journal of Disability Research* 11(2): 97-110.

Ravneberg, B. (2010) "De rette tekniske hjelpemidlene" s. 197-212 i Tøssebro, J. (red.) *Funksjonshemming: politikk, arbeidsliv og hverdagsliv* Oslo: Universitetsforlaget

Report No. 17 to the Storting (2006) *An Information Society for All* Oslo: Norwegian Ministry of Government Administration and Reform

Sassi, S. (2005) `Cultural differentiation or social segregation? Four approaches to the digital divide` *New Media & Society* 7(5): 684-700.

Seymour, W. (2005) "ICTs and disability: Exploring the human dimensions of technological engagement" *Technology and Disability* 17:195-204.

Storsul, T., Arnseth, H. C., Bucher, T., Enli, G., Hontvedt, M., Kløvstad, V. & Maasø, A. (2008) *Nye nettfenomener Staten og delekulturen* (New Internet phenomenon The state and the sharing culture) Oslo: Institute for Media and Communication

Strauss, A. og Corbin, J. (1998) *Basics of Qualitative research Techniques and Procedures for Developing Grounded Theory* London: Sage

Svendsen, E. (2010) *"Tal her" Formidling av avanserte kommunikasjonshjelpemidler og betjeningssystemer ("Speak here" Dissemination of advanced communication devices and operating systems)* Prosjektrapport, Trondheim: NAV Hjelpemiddelsentral Sør-Trøndelag

Söderström, S. (2009a) "Offline social ties and online use of computers: A study of disabled youth and their use of ICT" *New Media & Society* 11(2): 709-727

Söderström, S. (2009b) "The significance of ICT in disabled youth's identity negotiations" *Scandinavian Journal of Disability Research* 11(2): 131-144.

Söderström, S. & Ytterhus, B. (2010) "The use and non-use of assistive technologies from the world of information and communication technology by visually impaired young people: a walk on the tightrope of peer- inclusion" *Disability & Society* 25(3): 303-315

Tøssebro, J., Engan, E. & Ytterhus, B. (2006). "En inkluderende skole?" ("An inclusive school?) in J. Tøssebro & B. Ytterhus (eds.) *Funksjonshemmede barn i familie og skole – idealer og hverdagspraksis (Disabled children in families and schools – ideals and everyday practices)* Oslo: Gyldendal akademisk

Tøssebro, J. & Lundeby, H. (2002) *Å vokse opp med funksjonshemming (To grow up with disability)* Oslo: Gyldendal akademisk.

UNESCO (1994) *The Salamanca Statement and Framework for Action on Special Needs Education* Paris: UNESCO

Vavik, L. et al., (2010) *Skolefagundersøkelsen 2009 (The study of scholl subjects 2009)* HSH-rapport 2010/1 Høgskolen i Stord/Haugesund

Wendelborg, C. (2010) *Barrierer mot deltakelse. Familier med barn og unge med nedsatt funksjonsevne (Barrier to participation. Families with children and youth with impairments)* Trondheim: NTNU Samfunnsforskning

Wendelborg, C. & Kvello, Ø. (2010) "Perceived social acceptance and peer intimacy among children with disabilities in regular schools in Norway" *Journal of Applied Research in Intellectual Disabilites* 23(2):143-153

Wendelborg, C. & Tøssebro, J. (2010) "Marginalisation processes in inclusive education in Norway – a longitudinal study of classroom participation *Disability & Society* 25(6): 701-714

Yu, L. (2006) `Understanding information inequality: Making sense of the literature of the information and digital divides` *Journal of Liberianship and Information Science* 38 (4) 229-252.

Wielandt, T., McKenna, K., Tooth, L. & Strong, J. (2006) "Factor that predict the post-charge use of recommended assistive technology (AT)" *Disability & Rehabilitation: Assistive Technology*1(1-2): 29-40.

Winance, M. (2006) "Trying Out the Wheelchair. The Mutual Shaping of People and Devices through Adjustment" *Science, Technology & Human Values* 31(1): 52-72

Part 2

Communication and Social Interaction

3

Inclusion Through the Internet of Things

Louis Coetzee and Guillaume Olivrin
CSIR Meraka Institute
South Africa

1. Introduction

Inclusion of Persons with Disabilities and the Aged in mainstream society has progressed significantly through the continuous development of assistive technologies, standards, policies and guidelines combined with the use of Information and Communications Technology (ICT). Technological advances utilising standards developed to enhance inclusion have lowered the barrier to access information, for example, the World Wide Web Consortium (W3C) Web Content Accessibility Guidelines (WCAG) (*Web Accessibility Initiative Guidelines and Techniques*, 2011). WCAG provides recommendations to make Web content more accessible. Furthermore, the United Nations's "Convention on the Rights of Persons with Disabilities" is mainstreaming disability (Secretariat for the Convention on the Rights of Persons with Disabilities, 2011). Through the Convention, the Rights of Persons with Disabilities are described in detail. Member countries of the Convention commit themselves to institute mechanisms (be it policies, laws or other measures) to secure the individual's rights as captured in the Convention.

Through application of a philosophy such as "Design for All" (Stephanidis et al., 1999), awareness is being raised of designing and constructing devices and environments to accommodate the broadest spectrum of users without any modifications.

The National Accessibility Programme in South Africa is an example of a disability-related initiative aimed at enhanced inclusion and access to information through the Internet and ICT (Coetzee et al., 2007; Coetzee, Olivrin & Viviers, 2009).

Taking cognisance of the above progress, the challenges in ensuring continued and improved inclusion for Persons with Disabilities and the Aged have not yet been overcome, even though implemented mechanisms and technologies are lowering the barrier to inclusion.

Technology is advancing at a rapid rate and does not necessarily follow design principles aimed at enhancing or ensuring inclusion. Along the same vein, society is changing the ways in which it operates, utilizing the technologies in new ways. The establishment of Web 2.0 (also known as the Social Web) has changed the way information is gathered and shared, which subsequently has impacted on Persons with Disabilities and the Aged and their inclusion. Entities such as Facebook and YouTube, to name a few, are changing our daily lives, redefining how we interface with our friends and colleagues, and how we are entertained. These technological and societal changes do not necessarily enhance or ensure inclusion of Persons with Disabilities and the Aged. The importance of ICT as enabler for inclusion remains indisputably clear.

One very recent technological trend which will impact on society (and thus Persons with Disabilities and the Aged) is that of the **Internet of Things** (IoT) (Chui et al., 2010; International Telecommunications Union, 2005; Seventh Framework Programme, 2011).

In the IoT vision, all devices are connected to the Internet, each one with a unique Internet Protocol (IP) address and unique services rendered. Devices communicate and interact with one another and also provide information to higher-level integrated decision-making services and applications. Intelligent actioning can be initiated from these higher-level services and applications. The scale of this expanded Internet is unprecedented, with billions of devices connected and with masses of data generated. Value is created through the interpretation and analysis of the generated data and the resulting environmental actioning based on the analysis outcomes.

IoT is becoming a reality through a number of technological drivers: device processing and storage power are increasing; technology is becoming smaller while more sensors and actuators are being integrated; connectivity is improving – all facilitated through compliance to Internet related standards (e.g. IPv6). This provides an improved ability to connect, receive information, sense and act. IoT application spans a wide range of domains and areas. According to Vermesan et al. (Vermesan et al., 2011) the creation of smart environments/spaces is a major objective:

> "The major objectives for IoT are the creation of smart environments/spaces and self-aware things (for example: smart transport, products, cities, buildings, rural areas, energy, health, living, etc.) for climate, food, energy, mobility, digital society and health applications..."

As these connected devices become more integrated in our daily lives, so will the reality of the IoT vision. However, IoT's future role with regard to inclusion in smart environments is still unclear. Will it introduce more barriers or will it become an enabler? Given the migration of people towards urban areas, more people are living in larger cities. It is recognized that the urban areas need to provide enhanced services to its citizens, thus the drive to smart environments. These smart environments should be created to ensure inclusion of the broadest grouping possible, thereby creating an **enabling smart environment**.

The precise form and function of how IoT can break the accessibility barriers are not known yet. What is known is that inclusive design needs to be a fundamental element in the creation of IoT-enabled smart environments. Adopting a philosophy of creating an enabling environment through IoT, which embodies inclusiveness rather than just a smart environment, will go a long way towards ensuring inclusion in our technological futures.

This chapter casts an eye into the future to sketch inclusion scenarios (and ask relevant questions to ensure inclusion in future) through IoT. A brief overview of applicable research and development is presented in Section 2. Section 3 introduces Internet of Things. Smart environments and its application in our future society are presented in Section 4. Section 5 describes a methodology used and the results obtained, which allows for the progression of a smart environment into an enabling environment geared towards inclusion. As presented in Section 5, Section 6 highlights the specific needs and subsequent IoT-related services to ensure the evolution from *smart* to *enabled*. The usage of technological solutions can support an individual in exercising his rights as expressed in the United Nations's Convention on the Rights of Persons with Disabilities. Section 7 presents some of the convention's articles through which IoT can enhance or limit an individual's rights. Concluding remarks are presented in Section 8.

2. Literature

The role of smart environments (and their real-world instantiations – cities, houses, offices) in empowering the citizen, is the focal point of many different research projects (past and present). Research has progressed to the point where smart cities are purposefully being built or existing cities retrofitted. Examples include Amsterdam (The Netherlands), New Songdo City (South Korea), Lavasa (India), Skolkovo (Russia), Taihu New City (Wuxi, China), Dubuque (Iowa, USA) and several others.

Several reports and research outputs have been published regarding these smart environments. These publications include Berthon and Guittat's analysis of the rise of the intelligent city with the focus on managing resources in a sustainable way and the creation of an attractive economic and social environment allowing for interaction between citizens, companies and governments (Berthon & Guittat, 2011); Hill et al. analyse how ICT can transform energy-intensive establishments into low-carbon future smart cities where residents are enabled to make better, more informed choices. (Hill et al., 2011); Hodgkinson views digitally enabled inclusion in two dimensions: inside-out/formal where authorities are building more efficient infrastructures and services and outside-in/emergent where individuals create initiatives in support of the citizen (e.g. urban action forums, volunteer networks and carpooling networks) (Hodgkinson, 2011); Green views smart cities as a collection of programmes and concepts addressing environmental sustainability, economic performance, community cohesion and efficiency of operations (Green, 2011).

The role of future technologies and specifically that of the Future Internet in Smart Cities is discussed in the literature. Hernández-Muñoz et al. see the Future Internet as having the required building blocks (that of the Internet of Things and the Internet of Services) for creating an open innovation platform which can manage the various heterogeneous information sources, devices and data as associated with a future smart city (Hernández-Muñoz et al., 2011). Schaffers et al. investigate the role of the Future Internet experimentally-driven research and projects in the domain of Living Labs in smart cities (Schaffers et al., 2011).

IoT and its application in the real world is a very new research area. Large-scale projects, such as those funded by the European Commission, for example, the Internet of Things Architecture and Initiative (*IOT-A: Internet of Things Architecture*, 2011; *IOT-I: Internet of Things Initiative*, 2011), are still in their early stages. It has been seen that IoT can reduce the complexity and allow for the creation of smart environments (e.g. SmartSantander (*SmartSantander*, 2011)). Initial research in using IoT to ensure inclusion in specific environments is now appearing (Dohr et al., 2010; McCullagh & Augusto, 2011), but as yet has not made big inroads.

The following section introduces IoT and its various application categories, opportunities and challenges.

3. Internet of Things

Technological advances are creating new opportunities, services, mechanisms and business models to improve the quality of an individual's life. The advances are impacting on the ways people interface, interact and relate to the environment – be it the built, digital or social environment. One such technology redefining our world has been the Internet.

The Internet is an evolving entity which is becoming more important as broadband connectivity becomes ubiquitous. As depicted in Figure 1, the Internet started as the *Internet of Computers*. It has evolved into the *Internet of People* and is progressing towards the *Internet of Things*.

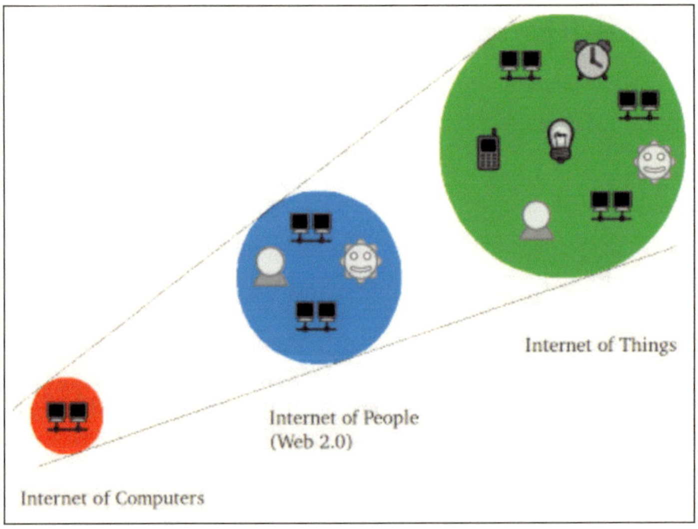

Fig. 1. Internet Evolution

The *Internet of Computers* provided a global network and a connected platform to run a variety of services. The World Wide Web is an example of an Internet service which was built on top of the original platform. Recently, more advanced services and technologies have changed the *Internet of Computers* into the *Internet of People*, known as Web 2.0 or the Social Web. In the Social Web, people are the main creators and consumers of content. Social networking such as *Facebook* and content sharing through a site such as *YouTube* illustrate that the Internet has matured and has evolved from the *Internet of Computers*.

A new technological wave is on the Internet horizon, that of the *Internet of Things* (Chui et al., 2010; Coetzee & Eksteen, 2011; Fleisch, 2010; International Telecommunications Union, 2005). Cheap and ubiquitous broadband Internet connectivity is creating an "*always on from anywhere*" opportunity. Combined with increases in device processing capability and storage capacity and associated reduction in the size of devices and adoption of standards, an additional dimension of *any thing* providing and using *any service* becomes a reality. These *things* often incorporate a variety of sensors, thus creating unprecedented masses of data to store and process. The resulting processed output can be fed back to the thing or other things in the surrounding environment where appropriate actioning can be instigated through actuators.

The definition of a *thing* is wide and includes a large variety of physical elements which include:

- Personal objects we carry (smart phones, tables, digital cameras, etc.).
- Elements and appliances in our environments (home, vehicle, work, urban or other).
- Objects fitted with tags (e.g. RFID (Bonsor & Keener, 2010), NFC (Nosowitz, 2011) or QR (O'Brien, 2010)) which are connected to the Internet (and has a cyber-representation) through a gateway device.
- Objects fitted with tiny computers – *smart* things.

Based on this view, an enormous number (billions) of devices and things will be Internet connected, each providing data, information and some even services. Once accurate information about status, location and identity is available at a higher system level, an opportunity is created for smarter decision making and action taking. Interactions can be between *thing* and *thing* (also known as machine-to-machine) as well as *thing* and *person* (machine-to-person and person-to-machine). In this phase of the Internet evolution, things have surpassed people as the main creators and consumers of data and content.

Illustrated in Figure 2 is a layered view of the IoT. The physical is connected via the Internet to middleware platforms (which provide the "standard" services such as *naming*, *discovery* and *security*). Services (to be consumed by other services or applications) and applications are built on top of the middleware. The inherent value and potential for impact, which can be obtained through IoT, increase when rising to the higher level services and applications and when these services are combined into *super* services and applications. The combination of services becomes a reality through the open architecture and philosophy of IoT.

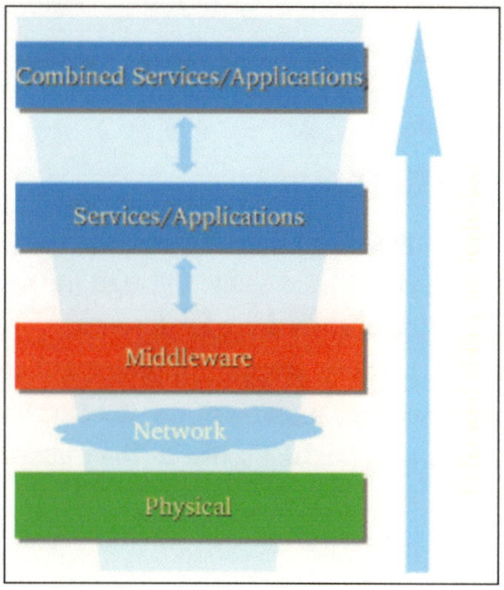

Fig. 2. Internet of Things Architectural View

Figure 3 depicts the fundamental elements in the IoT eco-system. Physical elements can be tagged and through the use of a gateway device (such as a mobile phone with a near field communications scanner) be linked to its cyber-representation on the Internet. Similarly, very small elements which are networked and which contains sensors to communicate sensed values via the network, are seen as part of the IoT. Traditionally, sensor networks operated in isolation. Through IoT, a multitude of sensor networks can be connected to the broader Internet, each of these smaller networks contributing information to a much larger audience. A fourth element in the IoT is that of *smart things*. These smart devices typically include mobile technologies such as smart phones, tablets and other traditional computing devices, as well as non-traditional elements such as household appliances that have computational power, network connectivity and sensors and actuators. As all the above *things* are network connected, the data they provide can be integrated and fused for use in integrative services and applications. These services, as well as the connected things, impact on the environment (people, industry and society).

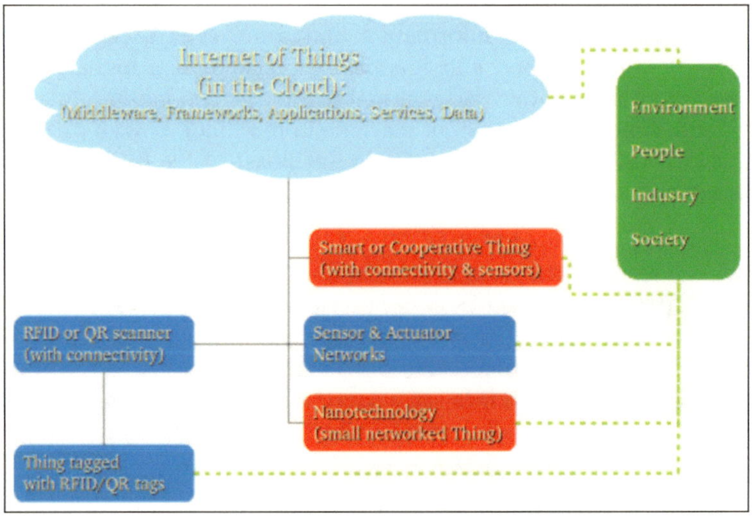

Fig. 3. Internet of Things Components

A vision for the IoT can be expressed as follows:

- All things become part of the Internet (or have a cyber-representation in the Internet cloud).
- Each thing is uniquely identified and accessible through the network.
- The position, identity and status of each thing are known.
- Each thing can sense, actuate, identify, interact, interface and communicate.
- Some things have intelligent services making sense of its localized data.
- Internet-based services exist, which make sense of the masses of data received from the connected things.
- Everything will be connected to everything else: Any **place**, any **time**, any **thing**, any **one**, on any **network**, using any **service**.
- Ultimately the IoT adds value to our world.

Figure 4 presents this vision in a graphical format.

Most of the enabling technologies for the IoT already exist (some not having optimal form or function yet, but able to contribute to the IoT). Based on this, the key driver for the adoption of IoT lies in the applications and new ways of solving existing challenges.

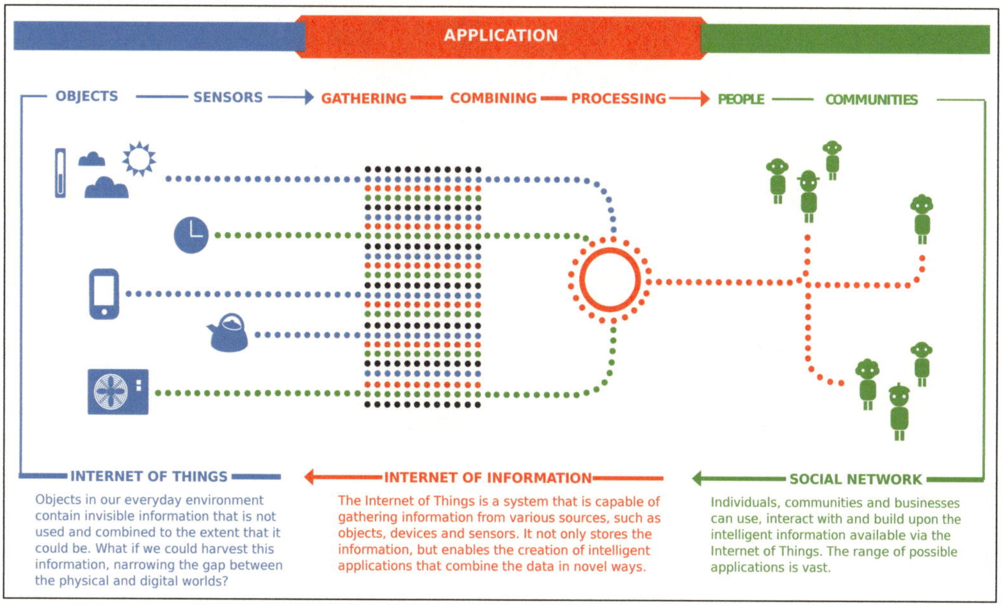

Fig. 4. Internet of Things Vision

3.1 Applications

IoT concepts have been demonstrated in a variety of domains: logistics, transport and asset tracking, smart environments (homes, buildings and infrastructure), health, energy, defence, retail and agriculture (Sundmaeker et al., 2010). IoT has the potential to significantly influence all facets of society.

According to Chui et al. (Chui et al., 2010) IoT applications fall into a number of broad categories. Figure 5 provides a graphical view of these application categories.

Chui et al. define two broad categories for IoT applications: *Information and Analysis* and secondly *Automation and Control*. Within each broad category, they further identify the following possible application of IoT concepts.

In Information and Analysis, decision-making services are enhanced by receiving better and more up-to-date information from networked elements in the environment, allowing for a more accurate analysis of the current status quo. This category applies to tracking (e.g. products in a logistics value chain), situational awareness (e.g. sensors in infrastructure or environmental conditions such as temperature and moisture) through real-time event feedback and sensor-driven decision analytics, which introduce concepts revolving around longer-term, more complex planning and decision making such as user shopping patterns in malls and stores.

Automation and Control implies acting on outputs as received from processed data and analysis. Process optimisation in industry is a promising application. A typical example would be where sensors measure the composition of a chemical compound, communicate it to a central service, where-after the service analyses and accordingly adjusts actuators to fine tune the composition. Optimised resource consumption can potentially change usage

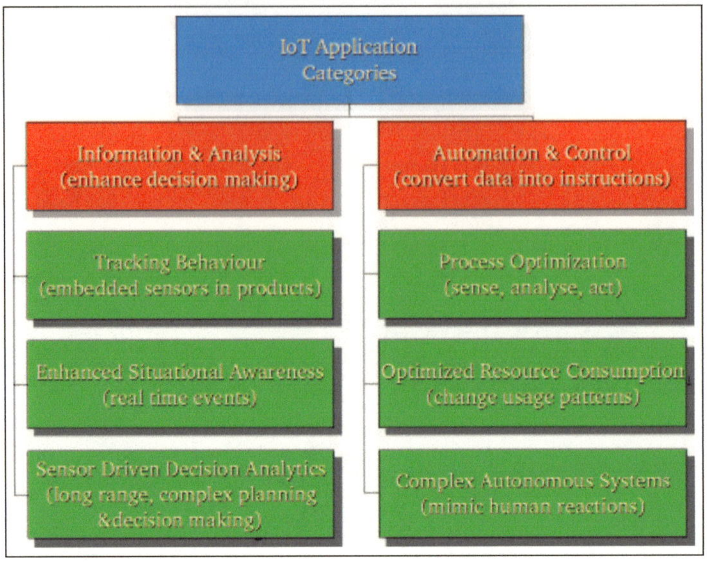

Fig. 5. Internet of Things Application Categorization (Chui et al., 2010)

patterns associated with scarce resources. Sensing and communicating the consumption of energy in households or data centres allow owners to adjust or load balance their usage to off-peak times with potentially lower costs. According to Chui, the real-time sensing of unpredictable conditions and the subsequent action taking based on those conditions, are a promising field of application. These types of applications mimic human behaviour (e.g. detecting an obstruction in front of a vehicle and then to initiate the appropriate evasive action) and are challenging to develop but holds promise for safety and security.

Fleish (Fleisch, 2010) provides a different view of the possibilities provided through the adoption of an IoT view.

According to Fleisch, IoT is relevant in every step in every value chain. He identifies seven main value drivers. The first four are based on value from machine-to-machine communication, while the last three create value with the integration of users. The drivers as identified by Fleisch are:

- **Simplified manual proximity trigger** – things can communicate their identity when they are moved into the sensing space of a sensor. Once the identity is known and communicated, a specific action or transaction can be triggered.
- **Automatic proximity trigger** – an action is triggered automatically when the physical distance of two things drops below (or passes) a threshold. The identity of the thing is known; when combined with the physical location and action, this allows for better processes.
- **Automatic sensor triggering** – a smart (or cooperative) thing can collect data via any type of sensor, including temperature, acceleration, orientation, vibration and humidity. The thing senses its condition and environment, and communicates the information which enables prompt (and global) decision making.

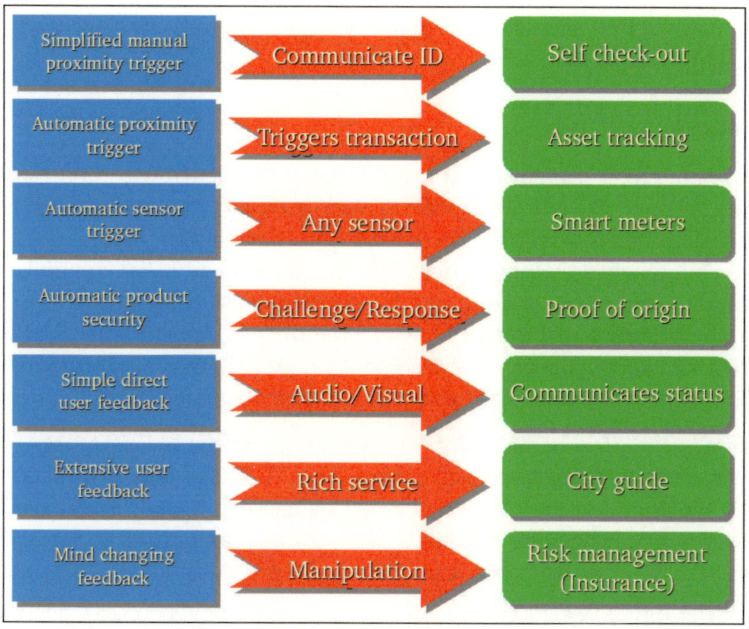

Fig. 6. Internet of Things Application Categorization (Fleisch, 2010)

- **Automatic product security** – a thing can provide derived security (information) based on the interaction between the thing and its cyberspace representation (e.g. a QR-code containing a specific URL pointing to relevant information).
- **Simple and direct user feedback** – things can incorporate simple mechanisms to provide feedback to a human present in the environment. Often these feedback mechanisms are in the form of audio (audible beep) or visual (flashing light) signals.
- **Extensive user feedback** – things can provide rich services to a human (often the thing is linked to a service in cyberspace through a gateway device such as a smart phone). Augmented product information is a good example of extensive user feedback.
- **Mind changing feedback** – the combination of real world and cyberspace might generate a new level of changing behaviours in people. One possibility is changing the driving behaviour as sensors in the vehicle communicate driving patterns to an outside agency.

IoT concepts and methodologies are being used more frequently in the creation of smart environments. The European Council's Framework Seven Programme is supporting a large-scale IoT project with *smart environments* as focus. The FP7 *Smart Santander* project hosted in the Spanish city Santander is positioned to be a large-scale IoT experimental research facility in support of typical applications and services for a smart city (*SmartSantander*, 2011). In Sweden, the *Sense Smart City* project (*Sense Smart City*, 2011) is researching ICT solutions for *smarter* urban cities and areas. Similarly, the city of Amsterdam is running the *Amsterdam Smart City* project (*Amsterdam Smart City*, 2011). Dohr (Dohr et al., 2010) utilises IoT in an ambient assisted living context to provide services for the elderly. McCullagh and Augusto investigate the potential of IoT to monitor health and wellness (McCullagh & Augusto, 2011).

Regardless of the categorization, it is clear that IoT can enhance inclusion of Persons with Disabilities and the Aged, or just as easily exclude Persons with Disabilities and the Aged.

IoT can easily impede the rights of individuals. Section 7 analyses the potential impact on the rights of the individual further.

3.2 Opportunities and challenges

Not all aspects of IoT have been resolved to the point of seamless integration. Some challenges remain. Most significant of these are aspects related to *privacy*, *trust* and *security*. Some of these aspects have been partially answered by some applications of the Internet of People, such as the Social Web, but the introduction of objects to this Internet adds the complexity of resource sharing, attribution and usage management. In the IoT world the question of **who** can see and act on **what** remains unanswered. How can the rights of the individual be ensured? The *societal* and *ethical* elements that will be created through IoT, needs to be explored. Similarly, *business models*, *governance* and *policies* still need to be resolved.

Many technological platforms and solutions will make up the IoT. Ensuring *interoperability* through appropriate *standardisation* is a high priority. The envisioned scale and complexity introduced by the large number of participating elements is an enormous challenge. How can the *robustness* of solutions be ensured? IoT solutions and applications will face a data deluge. How will the data from billions of *things* be processed, stored and maintained for future generations where appropriate?

Smart environments are becoming more prominent. The following section presents smart cities and smart homes and highlight the potential of IoT as enabler for smart environments.

4. Smart environments

The concept of *smart environments* – be it city, urban or home – has been the focus of many different research projects and implementations (Hodgkinson, 2011; Robles & Kim, 2010; Taylor et al., 2007). Multiple definitions of *smart environments* exist. A common theme in the definitions is how ICT can enable the delivery of services to the people in the environment, leading to improvements in *service delivery*, *governance*, *transport*, *sustainability*, *energy* and *innovation*.

Schaffers et al. provide a description of the many concepts associated with the smart environment (Schaffers et al., 2011):

"**Cyber cities** from cyberspace, cybernetics, governance and control spaces based on information feedback, city governance; but also meaning the negative/dark sides of cyberspace, cybercrime, tracking, identification, military control over cities.

Digital cities from digital representation of cities, virtual cities, digital metaphor of cities, cities of avatars, second life cities, simulation (sim) city.

Intelligent cities from the new intelligence of cities, collective intelligence of citizens, distributed intelligence, crowdsourcing, online collaboration, broadband for innovation, social capital of cities, collaborative learning and innovation, people-driven innovation.

Smart cities from smart phones, mobile devices, sensors, embedded systems, smart environments, smart meters, and instrumentation sustaining the intelligence of cities..."

Fig. 7. Transport service in smart cities: virtual descriptions of the environments provide both simulation and documentation of the service

In most cases, the creation of such an environment has been achieved through the creation of an ICT vertical solution. These solutions typically do not allow for interoperability, which limits the potential benefits that can be obtained. The promise of IoT to create an open, standards-based, integrative solution to integrate the physical world into cyberspace and provide services that can enable smart environments, has been recognized. Vermesan et al. (Vermesan et al., 2011) recognize the enabling role IoT can play in the creation of smart environments. Similarly, Hernández-Muñoz et al. (Hernández-Muñoz et al., 2011) see the European Commission's Future Internet research initiative (which has as building blocks IoT and the Internet of Services) as a mechanism to avoid the current vertical technological islands typically associated with smart cities.

4.1 Smart cities

There is a clear progression in terms of sophistication and integration of smart cities. One of the aims in this progression is to allow for better decision making by citizens as well as enhancing resource utilization (e.g. energy and water) and lowering the city's carbon footprint. Giffinger et al. define the characteristics of a smart city as (Giffinger et al., 2007): Smart Economy; Smart People; Smart Governance; Smart Mobility; Smart Environment and Smart Living.

ICT as enabler realizes the following progression:

- ICT as a support function.
- ICT providing information to citizens.
- ICT providing access to real-time data which aid decision making.
- ICT acting on the information to control the environment.

Taking the above into account, the following categorization of *smart cities* can be created:

- **Ubiquitous infrastructures** – broadband providing always on connectivity for both devices and people.

- **Rich in information** – an abundance of sensors to determine the state of the environment (temperature, humidity, traffic, open parking, etc.) and access to data.
- **Smart services** – advanced services from city to citizen and citizen to citizen, which utilize the available information as collected through the various infrastructure elements.
- **Action taking** – an abundance of actuators and other feedback mechanisms through which the environment is influenced and controlled.

4.2 Smart homes

Aldrich provides a good categorization of *smart homes*. He provides the following breakdown (Aldrich, 2003):

- "**Homes which contain intelligent objects** – homes contain single, standalone appliances and objects which function in an intelligent manner.
- **Homes which contain intelligent, communicating objects** – homes contain appliances and objects which function intelligently in their own right and which also exchange information between one another to increase functionality.
- **Connected homes** – homes have internal and external networks, allowing interactive and remote control of systems, as well as access to services and information, both from within and beyond the home.
- **Learning homes** – patterns of activity in the homes are recorded and the accumulated data are used to anticipate users' needs and to control the technology accordingly.
- **Attentive homes** – the activity and location of people and objects within the homes are constantly registered, and this information is used to control technology in anticipation of the occupants' needs..."

Several IoT application categories have been suggested (as described in Section 3.1). These are depicted in Figures 5 and 6. The above categorization of smart homes and cities is aligned and applicable with the IoT application categories, thus confirming IoT as enabler for these environments.

With more connected objects (with sensors and actuators) in an environment, services incorporating those objects become part of the landscape. As depicted in Figure 2, the value provided by IoT increases by composing services into higher-level services. Driven by IoT's fundamental interoperability characteristic, the basic services in an environment can be composed into higher-level services, with the subsequent increase in potential impact. These high-level services allow for enhanced *learning* and *attentiveness* and can provide for "*mind changing feedback*", "*complex planning and decision making*" and "*mimicking human reactions*".

Smart environments do not by default ensure inclusion. An enabling smart environment geared towards inclusion is required. Section 5 presents a methodology used to extract needs of Persons with Disabilities and the Aged. It highlights technological gaps and presents arguments supporting IoT as enabler for enabling environments.

5. Inclusive and enabling environments

As stated by Vermesan et al. (Vermesan et al., 2011), IoT has as its major objective the creation of smart environments. However, IoT also has the potential to evolve the smart environment into an enabling environment.

An *enabling environment* is an environment – either physical or virtual – that is designed or augmented in such a way that everyone, irrespective of disability or age, experiences equal participation and faces no barriers to activities, integration and independence.

In order to meet the objectives set by this definition, a research project called "Enabling Environments" was conducted in 2006 at the CSIR Meraka Institute, in Pretoria, South Africa (Macagnano, 2008; Williams et al., 2008) to determine possible technical interventions applicable to an urban environment and in a developing country.

The research sought to answer the following question:

> How do you integrate People, Technologies and the Environment into an interactive information and communication system, where the physical and virtual infrastructures become intelligent, connected and able to respond to human psychological and physical needs, in an accessible and usable manner?

In order to answer this question, various methodologies were employed, mostly to understand the environment and determine the needs of Persons with Disabilities and the Aged. The approach was people-centric, value-based, and hinted at technical solutions that were fields for future research.

5.1 A people-centric view of the Enabling Environment

The 2006 Enabling Environments research project approached the research challenge as a triad: Environment, People and Technology. As such, it conducted environmental audits, people surveys and workshops, and applied some future thinking and idea-generation techniques to identify candidate technologies with the potential to be a part of the solution.

The first method used to understand the context of an environment is called Future-Thinking (Hietanen et al., 2011). A workshop was organized, which included Persons with Disabilities and the Aged as members of the team to identify actors, customers, needs, products and services, environments, transformation processes, values, obstacles as well as political, economic, social, technological and ecological "drivers". Once a contextual picture is drawn, it is possible to generate interesting people-centred, value-driven and service-orientated scenarios. In these typical scenarios, there would be challenges that technology may be able to solve. Often, there was no simple technological answer to an environmental problem for a specific disability as each circumstance is different and makes it difficult to conceive a universal design that works for all.

The outcomes of the future study lead the team to select a physical location (Church Square, Pretoria, in South Africa) in which environmental audits and interviews with Persons with Disabilities and the Aged were conducted. Several challenges were readily identified:

- **Mobility** – the inability to use transport infrastructure and access places.
- **Wayfinding** – difficulty of finding a route usable with a wheelchair, or finding the easiest path with reasonable places to rest, or finding the path with the least risk (avoiding construction areas, for instance).
- **Communication** – the language and cultural barriers related to asking for information, or help, and knowing the protocols and the terms of the exchange.
- **Information** – Inaccessible information: signs which are not localised, transport routes and timetables, which are not physically usable (too high on a wall, print fonts too small,

small crowded space with one resource providing information). This also includes people experiencing difficulties to read (illiteracy) or see (visual impairment), and the general lack of information that is useful to the individual, versus the noise and over-abundance of information provided in the environment (warnings, documentation, recommendations and guidelines).

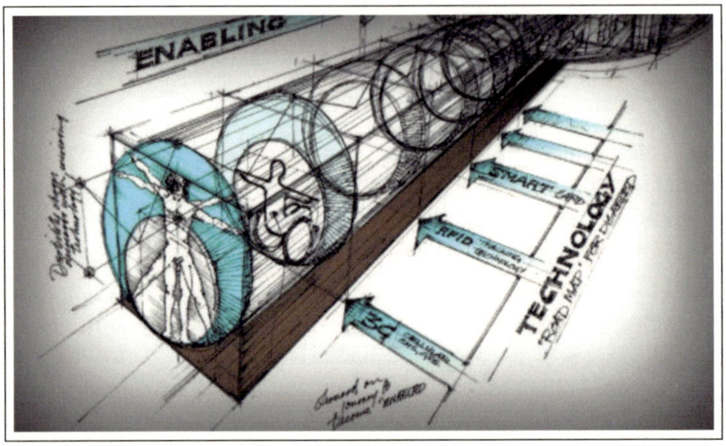

Fig. 8. User-centric and universal design: the environment must enable any person without assistive technology, at the same time, services must be adapted to each person

Following the initial Future-Thinking study and the field surveys, the team held a persona and scenario workshop to extract a full account of those challenges that were encountered by a blind person, a Deaf person, a quadriplegic wheel-chair user and an elderly person in the city centre (Williams et al., 2008). For each persona, a different scenario best illustrated the challenges that were met. The scenarios used were:

- Catching a taxi and finding your way to your destination as a blind person.
- Communicating in a multilingual environment as a Deaf person.
- Crossing public spaces in a wheelchair.
- Finding help and reasonable places to rest as an elderly person.

These people-centric studies form the basis of our understanding of what technology should do and help in identifying gaps, specifically in a built environment such as a city (public places) and a room (office and home). They showed that there were technological gaps that had to be addressed in the environment, in a universal manner to benefit all. The studies further showed that an enabling environment had to be envisioned for individuals, and provide solutions addressing each of their use cases and intent in the environment. These two aspects were difficult to reconcile and it was indeed very difficult to provide one unified solution for all these individual and personalised needs based on specific assistive technologies. Instead, the problem has to be reformulated: **How does an individual create and use his own assistive service from the various devices present in the environment?**

5.2 Technological gaps

The specific technical solutions that were originally suggested in response to the persona and scenarios challenges relied more on providing specific assistive technologies to each persona to break down their environmental barriers and less on a universal design which achieves the same results through the interconnection of many things and improvements in the environments themselves.

The above research on enabling environments concluded that the solution lay in furthering the multi-disciplinary research and solving the following technical challenges:

- **Multiple Mobile Sensor Technology** – for the environment to become aware of its own state.
- **Universal Design, Accessibility and Usability** – for the environment to be usable by all.
- **User Profiling, Social Engineering, Social Research** – to know the context and profile of a person.
- **Context Awareness, World Modelling** – to model the context and environment with a virtual system.
- **Artificial Intelligent Reasoning** – to learn and reason about data, context, state, events and people.
- **Software Engineering and Computational Issues** – to computationally manage a connected, integrated intelligent environment.
- **Assistive Technologies and Human Computer Interaction** – to interact and engage with the enabled environment.

These are the premises for an enabling environment, in which the technical stack will need to incorporate a world of sensors and provide personalized services in a user-centric manner.

After reviewing the technical challenges and gaps, which were identified in the Enabling Environment project, IoT seems to provide possible technical solutions. Typically, IoT is an environment itself with practical implementation for the concepts of ambient and ubiquitous intelligence. From this perspective, there is less emphasis on smart devices and assistive technologies ("technological interventions" in the environment or with the people) and greater emphasis on existing environmental things and the possible utilization scenarios individuals can create for themselves.

This approach is more technology-centric at first, enabling the things that make up the environment, before their combined data and complementary value provide all people with "*enabling smarts*". The question remains: what are the technologies that must be introduced as a standard to provide enabling smarts that people can use? IoT provides such solutions which integrates the idea of disaggregated assistive services for all.

5.3 Internet of Things as the enabler

IoT provides an integrated solution which addresses four of the technological gaps expressed above:

- How to make the environment aware of its own state.
- How to model the context in a virtual system.
- How to provide intelligent services based on learning and reasoning.
- How to computationally manage this connected and integrated environment.

IoT is able to provide some "*enabling smarts*" by finding value in the combination of many things that are already in the environment, and possibly on the person. IoT is a mechanism through which Weiser's original *Ubiquitous Computing* vision (Weiser, 1991) and its extension, the *Ambient Intelligence* vision (Gill, 2008), can be realized.

IoT therefore provides an implementation framework for *"enabling smarts"* such as:

- **Augmented Reality** – IoT provides the means to virtualise the physical environment from all the sensors and actuators, which are present. It makes possible an alternative rendition of the environment by filtering, reorganizing sensory channels and changing the modality of the information altogether.
- **Disaggregated Computing** – rather than one powerful assistive device (often *attached* to the person), use the various capabilities of the many environmental things to provide a more reliable service.
- **Invisible Computing** – IoT makes it possible to focus on the task at hand rather than on the assistive technology to bridge the gap.

A smart environment becomes an enabling environment when the combination and processing of information benefits the person through the following:

- **Identify and Personalise** – by knowing the person's needs and adapting accordingly.
- **Track and Sense** – by knowing the person's location and understanding the intention and objective.
- **Inform** – by telling the person what he needs to know and by describing his options in specific scenarios.
- **Consult and Act** – by asking the person for feedback and then instigating changes in the environment beneficial to the person.

These classes of services closely match the application areas previously described for IoT (see Section 3.1). Enabling the things that already make up the environment means that people will find themselves empowered in an environment that is familiar and more conducive to tasks they want to execute.

5.4 Enabling Internet of Things applications

In the preceding sections, the generic applications and value of IoT as enabler have been discussed. These need to be translated into real-world applications that can be implemented. How can the described scenarios be addressed through IoT?

Catching a taxi and finding your way to your destination as a blind person. As a first priority, the blind person needs to be aware of his own location. This can be obtained through a smart mobile device (e.g. smart phone) with a GPS sensor. Being in a fully connected environment he can initiate a *call* to a taxi service, which contains his GPS coordinates. A taxi (which is connected to the IoT) will be messaged and instructed to drive to the closest taxi rank to the received GPS coordinates. The blind caller will be routed to the appropriate taxi rank via embedded indicators in the environment. These indicators might be talking side walks and traffic lights, which have been notified of the blind person's route to that appropriate taxi rank. A traffic light will sense that the person is approaching (by sensing the person's RFID tag which is embedded in his smart phone) and will regulate the traffic (e.g. stop the flow) as the person crosses the road. Alternatively, the indicators can be audio queues from the side walk, which are activated as he moves towards the taxi rank. When he is approaching the taxi, the audio queues will guide the person towards it.

Fig. 9. Scenario for a blind man finding and catching a taxi: the presence of smart technology in the environment can enable this service

Communicating in a multilingual environment as a Deaf person. The environment is fitted with a multitude of Internet-connected displays. The environment also has a multitude of RFID or other near field communication (NFC) scanners. The Deaf person will be in possession of a tag (either stand alone and embedded in his clothes or part of a smart hand-held device such as a smart phone) which is continuously scanned by the environment. As his identity and location and subsequent preferences are known in cyberspace, the connected displays in the person's vicinity can sense his presence and adapt their signage to the appropriate modality (e.g. Sign Language in the specific dialect instead of normal written text).

Crossing public spaces in a wheelchair. The wheelchair user's current location is known, as sensed in the environment (e.g. RFID scanners able to scan the RFID chip as embedded in the wheel chair). The person communicates his target location by speaking (microphones in the environment record the audio signal and cloud-based services perform automatic speech recognition). The routing service calculates an optimal route, taking into account the current state of the environment (e.g. construction in an area, which would prevent a wheelchair from passing through). The appropriate route is communicated to him via connected displays en route (e.g. routing a person to avoid stairs and steps, but rather to ramps and lifts). As the wheelchair user moves, his position is tracked, with those displays in his vicinity presenting him with updated routing information for the environment.

Finding help and rest as an elderly person. The IoT environment is aware of a specific person (and also the person's profile and abilities). An always on and sensed environment, listens to the person, and responds when the person asks for help (microphones, automatic speech recognition or the press of a *help* button on an interactive display, or through accessing a city service from his smart mobile device). The environment, through the signalling, can guide the person to a place of rest (e.g. restaurant with open tables, or closest public seating in the park).

Smart-enabling environments can guide and advise people accessing its space and using its services. Smart-enabling environments are aware of people and can optimize certain processes, taking into account all the people who require a service or access to a space, to ensure it adapts when appropriate to accommodate other people. In other words, it is possible to make some changes in signalisation and information in order to route and help people in making better choices that will benefit the majority, but also to benefit the few that are in need (lost, tired, agoraphobic, claustrophobic, clearing a path). An IoT-enabled environment can aid in the following ways:

- Information:
  - Access to personalized signage: environment communicates in appropriate modalities using appropriate devices/appliances.
- Communication:
  - Signalling for help: environment senses the need (audio, visual, signal from smart device) and responds appropriately.
  - Getting service: a specific service request is communicated to the environment (e.g. to access public transport), whereupon the environment responds.
  - Payment: in a connected environment, the person's identity is known and the link to a bank account accessible to the service is available. Auto payments based on the specific action takes place.
- Finding places:
  - Navigation: location, direction, orientation and wayfinding according to specific parameters (energy level, time-wise, crowd-wise) become possible. For example, motorized wheelchairs communicate their battery charge levels, location and in return, receive relevant information which takes the context into account.
  - Virtual presence in a real environment: accessing services and even walking in the park.
- Mobility:
  - Planning transport and routes: receiving information from the environment regarding the status of transport in the environment.
  - Access to transport information: getting on and off transport, knowing if a specific mode of transportation is accessible and usable for that person's specific needs.
  - Access to site information: determining accessibility of a site depending on the mode of transport (by foot, bicycle, train, bus, taxi or wheelchair).

From the above narrative, the role of IoT in enhancing inclusion in future environments (home or city) is clear. However, to enjoy the benefits, IoT-related challenges need to be resolved. Section 6 analyses these aspects.

6. Inclusion benefits through the Internet of Things

In the preceding sections, IoT, smart environments and the road to an enabling environment were described. IoT as technology can create *smart environments*. IoT connects smart objects within an environment, senses the environment and collates and processes the sensed data in higher-level services, with the result fed back to the environment whereupon actioning in the environment is initiated. In smart cities, the focus is on *service delivery*, *governance*, *transport*, *sustainability*, *energy* and *innovation*. In smart homes, the focus is on *information exchange*, *increased functionality*, *distance* and *remote control* and facilitating *learning* and *attentiveness*.

When IoT-based services and technologies, which allow for improvement in **mobility**, **wayfinding**, **communication** and **access to information**, become part of the standard suite of services in a smart environment, the door to inclusion and an enabling environment has been opened.

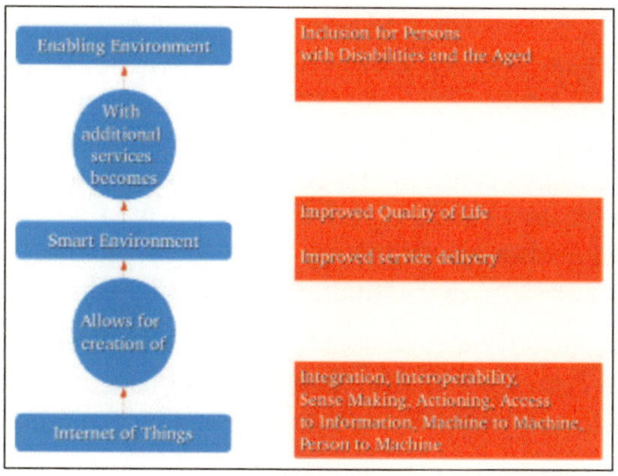

Fig. 10. Enabling Inclusion through IoT

Similarly, if IoT technologies empower and benefit the individual by **identifying** and **personalising**, **tracking** and **sensing**, **informing** and **consulting** and **acting** in appropriate and responsible ways, inclusion through enabling environments becomes a real possibility.

If it is fair to assume that the value of IoT emerges from an abundance of connected things, it is also important to consider how these things combine to provide services, and how it is possible to discover their features, understand their capabilities and usages, categorize them and filter them in reusable user libraries and taxonomies. Building IoT-enabling services from user scenarios is discussed in Section 6.1. The section identifies elements that make things usable in assistive scenarios.

The second benefits-versus-challenges aspect which IoT presents is discussed in Section 6.2: how is the Internet dimension of IoT an accessible and usable medium? Can it provide the required functionality for all individuals, irrespective of their abilities (and disabilities), learning style and preferences, language and level of literacy?

6.1 Enabling things

Section 5.4 provides some examples on how "enabling" applications of IoT can be conceived, based on documented people-environment scenarios. Service providers can follow the described methodology to find innovative applications and new usage of technologies and create new services for the city and in the house.

The real test, however, is whether a person with specific needs will be able to communicate his goals and receive a personalised service, by both publishing a request or an expression of his intent, and by piecing together his own assistive environment from the various things in his contextual environment.

In either case, enabling things making up an assistive environment require discovering and combining properties of the environment into a workable solution.

This entails various levels of intent (being part of the solution or making requests) and various levels of control (privacy and interaction).

In the most informative approach, a person states his goals by publishing a profile which describes abilities, languages and other particulars. This person expects some degree of support and initiative from the environment.

In a more conservative approach, the person builds an assistive solution without stating his intentions, i.e. without publishing a profile. The person will then actively create a custom assistive service as an IoT application made up of the surrounding enabling things. Of course, most practical assistive solutions will be somewhere in the middle ground: from the person's perspective "nothing about the person [must be done] without the person". From a technology perspective, neither "pushing" nor "pulling" services are an acceptable solution; the technology must be accessible and communicate and interact with the person on demand.

From an IoT perspective, this "human accessible" dimension introduces new challenges which need to be addressed to change the smart environment into an enabling one. Below are some human-to-thing conversational aspects and requirements for the things themselves:

- **Service discovery** – discovering existing services and things in the environment. To discover the service, the context of the request is important: the context comprises both the physical conditions and the virtual states. A person might only be virtually present in the environment, but would still want access to the real world context. The person may be physically present in a context, but seek information about another context virtually.
- **Taxonomy and classifications of things, functional and assistive usages** – can be used to classify the assistive services and IoT applications from authoritative sources (e.g. ICF, ISO 9999, c.f. Heerkens et al. (2011)) and user-made folksonomies (Web 2.0 tagging for instance).
- **Documenting and providing alternative human interfaces (languages, representations) and software APIs** – this is required to provide information on how IoT applications can be built and how their composition create assistive services. Interfaces and adapters are essential to providing both human control and contracts between things.
- **Providing a language in which services can be described as a combination of smart enabling things** – domain-specific descriptions and terms are combined and coordinated into valuable scenarios (e.g. Pentagruel (Drey et al., 2009)). Design patterns, best practices and example scenarios need to be conceived to create specific behaviours and services, for instance, the Sense/Compute/Control model proposed by Cassou et al. (2011).

The key aspect for an enabling environment is for the person to know how to discover solutions. The assumption is as follows: Whether the person is asking for a service, or controlling and making environmental adjustments and adaptations, there is a degree of control over both the physical environment and the virtual, connected one.

The next section discusses how enabling environments via IoT depend on the existing efforts in accessibility, usability and ergonomy of both the real and virtual environments.

6.2 Accessing the Internet of Things

Getting access to the IoT is a question of finding a thing, interface or an Internet protocol which is best suited to the person's needs and goals.

This presents the usual challenges of both ergonomy and environmental design, and of accessibility and usability of information systems.

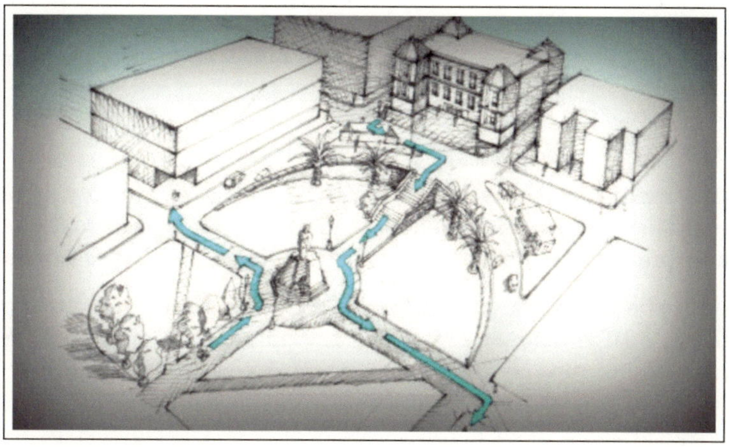

Fig. 11. Based on the person profile and preferences, find a path that's safe, that maximises rest and shade, and minimises effort

The vision is for both modes of interaction, either physical or virtual, to complement each others in such a way that the physical barriers are removed, and that the virtual world and information systems are less abstract and disembodied.

Technologies that currently make up the basic IoT inventory already introduce some physical barriers. Things in the environment (price-tags, schedules, menus, toilet availability, shop status – open or closed) may also be beyond the reach of the physical or sensory (barriers, height, effort, people), and therefore inaccessible to some people. Providing a virtual, connected representation means enhanced accessibility.

- Barcodes (1D, 2D) are popular and provide a unique identifier for an item and often, a reference link for the tagged object. Although barcodes are not connected, they can rely on connected scanners and cameras to become a thing of the Internet. Barcodes, however, are physically inaccessible for someone with a visual impairment, who will not notice nor be able to bring barcodes into the scanners' field of view. As a thing of the Internet, barcodes can be linked to and augmented by GPS coordinates or even alternative RFID or NFC technologies to provide a more accessible identification service. Relying on the community of people is another solution: the "eyes" of other people, through their smart devices, produce decoded information, e.g. GPS coordinates to an application of the IoT such as Thing-Memory (Coetzee, 2011) (a cyber representation and model of physical things). Blind people, supported by their GPS position and orientation, now have a way to explore the environment based on this cyber model of the environment.
- Public displays are popular for adverts, news and transport schedules and are in fact "connected" things. They pose the usual problems of accessibility for the blind and Deaf. Yet, as things of the Internet, they can make their information available through alternative formats such as text-to-speech for the blind and captions for the Deaf. Connected "Internet" TVs then become accessible through the IoT, beyond their physical limitations.
- Alarms, vehicle horns, sirens and other warning signals also ought to be delivered over a wider spectrum of modalities in an environment. If connected to the Internet, these things ultimately increase their reach, scope and impact not only through connected information

systems and social media, but also through other objects present in the environment that may physically change their state or behaviour to reflect the state of emergency or risk.

Accessibility is therefore increased through the additional modalities provided by connecting information systems and the physical environment.

The abundance of modalities means that information systems have to provide people with the most appropriate modality and format for the contextual information, based on the individual person's abilities, preferences, literacy and language (Coetzee, Viviers & Barnard, 2009).

People will need to be selective and define what constitutes their context, and swap between contexts to filter the large amount of information emerging from the IoT. People have to populate their environment with the things, sensors and services which they request from the IoT. Making these contexts familiar and safe is the biggest challenge that the IoT has to address.

Regarding usage and safety, the IoT is able to provide an important layer of helpful and critical information.

Firstly, it is possible for each thing, once identified, to offer people documentation, usage patterns, warnings and advice from other people. If connected, the thing can communicate an error or have its status tracked, verified and asserted by applications in the IoT.

Secondly, each thing can be assigned a level of safety and safety conditions to present people with a warning of possible dangers or potential misuse when entering its context. The inter-connectedness of things also means that the accumulated effects of things in a context can trigger a safety warning, such as electrical objects near water, magnets near magnetic media storage, etc.

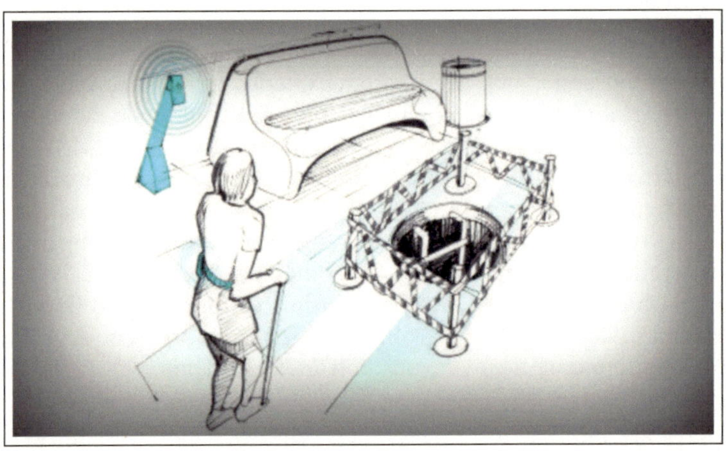

Fig. 12. Dangerous Things: the virtual layer provided by self-documenting, self-describing smart objects makes danger, that's beyond the reach of our senses, discernible

The final challenge for IoT in the assistive domain is that of trust and reliability: because the solution is made of disaggregated, heterogeneous components with varying quality, and relies on connectivity with most applications relying in turn on models and knowledge systems. The combination of all aspects means that trusting in an IoT application to provide an assistive service may prove too uncertain, unreliable and risky for people. Proving that these solutions can work reliably and reconfigure when aspects of the IoT change, forms yet another level of IoT integration dedicated to quality of service.

Persons with Disability and the Aged have rights as described in the Convention on the Rights of Persons with Disabilities. People are at risk of having their rights impeded through IoT-based technologies. The following section describes Articles applicable in an IoT environment and highlights the challenges in complying to all the stated rights.

7. United Nations's Convention on the Rights of Persons with Disabilities

The preceding sections described technologies and a methodology to extract user needs and application of technologies to address the identified user needs. However, cognisance of the wider context needs to be taken. Efforts to enhance inclusion is only part of the solution. All solutions implemented are framed by the Convention on the Rights of Persons with Disabilities.

The United Nations's *Convention on the Rights of Persons with Disabilities* (Secretariat for the Convention on the Rights of Persons with Disabilities, 2011) has been created to protect the rights and dignity of Persons with Disabilities. Article 1 states the Convention's purpose as follows:

> "The purpose of the present Convention is to promote, protect and ensure the full and equal enjoyment of all human rights and fundamental freedoms by all persons with disabilities, and to promote respect for their inherent dignity.
>
> Persons with disabilities include those who have long-term physical, mental, intellectual or sensory impairments, which in interaction with various barriers may hinder their full and effective participation in society on an equal basis with others..."

The Convention defines the actions that a Signatory (e.g. a Government) must execute to ensure that a person with disabilities can realize his right. The Convention addresses aspects that can be accomplished through IoT. However, IoT also introduces possibilities that can be detrimental to the individual's rights. The following articles have specific implications for IoT and Persons with Disabilities and the Aged:

Article 9 Accessibility "To enable persons with disabilities to live independently and participate fully in all aspects of life, States Parties shall take appropriate measures to ensure to persons with disabilities access, on an equal basis with others, to the physical environment, to transportation, to information and communications, including information and communications technologies and systems, and to other facilities and services open or provided to the public, both in urban and in rural areas..."

Article 17 Protecting the integrity of the person "Every person with disabilities has a right to respect for his or her physical and mental integrity on an equal basis with others..."

Article 19 Living independently and being included in the community "States Parties to the present Convention recognize the equal right of all persons with disabilities to live in the community, with choices equal to others..."

Article 20 Personal Mobility "States Parties shall take effective measures to ensure personal mobility with the greatest possible independence for persons with disabilities..."

Article 21 Freedom of expression and opinion, and access to information "States Parties shall take all appropriate measures to ensure that persons with disabilities can exercise the right to freedom of expression and opinion, including the freedom to seek, receive and impart information and ideas on an equal basis with others and through all forms of communication of their choice..."

Article 22 Respect for Privacy "No person with disabilities, regardless of place of residence or living arrangements, shall be subjected to arbitrary or unlawful interference with his or her privacy, family, home or correspondence or other types of communication or to unlawful attacks on his or her honour and reputation..."

Article 26 Habilitation and rehabilitation "States Parties shall take effective and appropriate measures, including through peer support, to enable persons with disabilities to attain and maintain maximum independence, full physical, mental, social and vocational ability, and full inclusion and participation in all aspects of life ..."

IoT services can easily impact on the above Articles and the associated rights. By complying with one article through IoT, the stated right of another article may be broken. An example is that of enhancing mobility for a visually impaired person. As stated in Section 5, *Mobility* and *Wayfinding* are key challenges for Persons with Disabilities and the Aged. Article 20 requires effective measures to ensure personal mobility. If an environment is fully enabled, with sensors positioned to keep track of the location of a specific person, a higher-level service can provide the required functionality to ease mobility and wayfinding. However, Article 22 requires *Respect for Privacy*. As the person's location is now known at all times, the privacy element can very easily be overridden. Because the person is tracked, and the information known at all times, the possibilities for abuse are rife.

Similarly, Article 21 addresses the individual's right to access information. IoT services can be developed to present information in the most appropriate modality (e.g. Sign Language for the Deaf). The manpower cost associated with creating Sign Language for information purposes may inhibit its creation. Information in other written languages (the localization process) is also expensive, but not as prohibitively so as that of Sign Language. In an African development context, there is the misconception that the Deaf are literate. This is not true, as literacy in itself is a big challenge for the Deaf. As a consequence, the Deaf can very easily be excluded in an IoT world, as Sign Language will not by default be seen as necessary.

Rights of individuals are important. Each technology that is introduced, should be measured against the Convention to ensure compliance and inclusion.

8. Conclusion

Inclusion of Persons with Disabilities and the Aged in our environments and societies has always been a challenge. Through policies and technologies, some progress is being made in improving inclusion. However, technology is evolving rapidly, with the consequence that while technology evolution can enhance inclusion, and it can also limit and impact negatively on inclusion. The Internet in general has improved inclusion through improved access to information and interaction with others (assuming that an appropriate assistive technology is used as is required).

IoT is the new Internet technology wave following that of the Social Web. Through the IoT, the Internet is extended into the physical world through connected objects or cyber-representations of the physical. The incorporation of the physical world into the Internet has the potential to further improve inclusion of Persons with Disabilities and the Aged. However, it can also impact negatively on inclusion. IoT has as its major objective the creation of smart environments and spaces and self-aware things. None of these ensures inclusion as a default. As IoT will become pervasive in our society and realized in environments such as smart cities, inclusion needs to be part of the fundamental design.

This chapter presents IoT, its typical applications and some of its challenges. It describes smart environments and the reasons for their increasing importance. It also links IoT as enabling technology for smart environments. It presents the methodology and the extracted generic challenges and technological gaps in creating an inclusive enabled environment. Through the matching of IoT application categories and the technological gaps, it is found that IoT is an enabler for four of the technology gaps in enabling environments: Through IoT, an environment can be made aware of its own state, the context modelled in a virtual system, intelligent services based on learning and reasoning provided, which are capable of computationally managing the connected and integrated environment. Furthermore, through IoT, services can be provided, which aid in identifying the individual and subsequently personalizing information and the environment, by tracking an individual and sensing his needs as well as the state of the environment, by informing the individual of options, by consulting with the individual and acting based on the retrieved information.

Even though IoT can assist in the creation of an inclusive enabling environment, cognisance must be taken of the implications IoT can have on the rights of Persons with Disabilities and the Aged. It is shown, through an analysis of the United Nations's Convention on the Rights of Persons with Disabilities, that the creation of services to comply with a specific Article, may compromise the right associated with another Article. Extreme caution thus needs to be exercised in the creation of an inclusive enabling environment.

The future holds the promise of greater inclusion through the upfront integration of IoT services and technologies. By raising awareness of the societal needs now, we can live in a more inclusive world tomorrow.

9. Acknowledgement

The authors acknowledge the contributions and inputs from the team members from the Ability Based Technology Interventions, National Accessibility Portal and Enabling Environments CSIR Meraka research projects. These include, but are not limited to Hina Patel, Quentin Williams, Ronell Alberts, Ilse Viviers, Harriet Easton, Ennio Macagnano, Bernard Smith and Gold Mametja. The authors acknowledges Johan Eksteen for his insightful contributions.

10. References

Aldrich, F. (2003). Smart Homes: Past, Present and Future, *in* R. Harper (ed.), *Inside the Smart Home*, Springer London, pp. 17–39. 10.1007/1-85233-854-7_2.
URL: *http://dx.doi.org/10.1007/1-85233-854-7_2*

Amsterdam Smart City (2011). Accessed December 25, 2011.
URL: *http://www.amsterdamsmartcity.nl*

Berthon, B. & Guittat, P. (2011). Rise of the Intelligent City, *Technical report*, Accenture.

Bonsor, K. & Keener, C. (2010). HowStuffWorks: "How RFID Works". Accessed December 25, 2011.
URL: *http://electronics.howstuffworks.com/gadgets/high-tech-gadgets/rfid.htm*

Cassou, D., Balland, E., Consel, C. & Lawall, J. (2011). Leveraging Software Architectures to Guide and Verify the Development of Sense/Compute/Control Applications, *ICSE'11: Proceedings of the 33rd International Conference on Software Engineering*, ACM, Honolulu, United States, pp. 431–440.
URL: *http://hal.inria.fr/inria-00537789/en/*

Chui, M., Löffler, M. & Roberts, R. (2010). The Internet of Things.
URL: *http://www.mckinseyquarterly.com/High_Tech/Hardware/The_Internet_of_Things_2538?gp=1*

Coetzee, L. (2011). Thing-Memory.
URL: *http://ioteg.meraka.csir.co.za/ioteg/contentView.seam?contentId=64*

Coetzee, L. & Eksteen, J. (2011). The Internet of Things – Promise for the future? An Introduction, *in* P. Cunningham & M. Cunningham (eds), *IST-Africa 2011 Conference Proceedings*, IST-Africa, IMC International Information Management Corporation, Gaberone, Botswana. ISBN: 978-1-905824-24-3.

Coetzee, L., Govender, N. & Viviers, I. (2007). The National Accessibility Portal: An Accessible Information Sharing Portal for the South African Disability Sector, *Proceedings of the 2007 International Cross-Disciplinary conference on Web Accessibility (W4A)*, ACM Press New York, NY, USA, Banff, Canada, pp. 44–53.

Coetzee, L., Olivrin, G. & Viviers, I. (2009). Accessibility Perspectives on Enabling South African Sign Language in the South African National Accessibility Portal, *Proceedings of the 2009 International Cross-Disciplinary Conference on Web Accessibility*, Madrid, Spain.

Coetzee, L., Viviers, I. & Barnard, E. (2009). Model Based Estimation for Multi-Modal User Interface Component Selection, *in* F. Nicolls (ed.), *Proceedings of the Twentieth Annual Symposium of the Pattern Recognition Association of South Africa*, Stellenbosch, South Africa, pp. 11–16. ISBN: 978-0-7992-2356-9.

Dohr, A., Modre-Osprian, R., Drobics, M., Hayn., D. & Schreier, G. (2010). The Internet of Things for Ambient Assisted Living, *Seventh International Conference on Information Technology*, IEEE.

Drey, Z., Mercadal, J. & Consel, C. (2009). A Taxonomy-Driven Approach to Visually Prototyping Pervasive Computing Applications, *1st IFIP Working Conference on Domain-Specific Languages*, Vol. 5658, Oxford, United Kingdom, pp. 78–99.
URL: *http://hal.inria.fr/inria-00403590/en*

Fleisch, E. (2010). Auto-ID Labs: What is the Internet of Things? - An Economic Perspective, *Technical report*, ETH Zurich / University of St. Gallen.
URL: *http://www.autoidlabs.org/uploads/media/AUTOIDLABS-WP-BIZAPP-53.pdf*

Giffinger, R., Fertner, C., Kramar, H., Kalasek, R., Pichler-Milanović, N. & Meijers, E. (2007). Smart cities – Ranking of European medium-sized cities.

Gill, J. (2008). Ambient Intelligence – Paving the way... Accessed December 25, 2011.
URL: *http://www.tiresias.org/cost219ter/ambient_intelligence/index.htm*

Green, J. (2011). Digital Urban Renewal. Reference Code: OT00037-004.

Heerkens, Y., T, B. & de Vrankrijker Kleijn-de, M. (2011). Classification and terminology of assistive products., *International Encyclopedia of Rehabilitation*.
URL: *http://cirrie.buffalo.edu/encyclopedia/en/article/265/*

Hernández-Muñoz, J., Vercher, J., Muñoz, L., Galache, J., Presser, M., Hernández Gómez, L. & Pettersson, J. (2011). Smart Cities at the Forefront of the Future Internet, *in* J. Domingue, A. Galis, A. Gavras, T. Zahariadis, D. Lambert, F. Cleary, P. Daras, S. Krco, H. Müller, M.-S. Li, H. Schaffers, V. Lotz, F. Alvarez, B. Stiller, S. Karnouskos, S. Avessta & M. Nilsson (eds), *The Future Internet*, Vol. 6656 of *Lecture Notes in Computer Science*, Springer Berlin Heidelberg, pp. 447–462.
URL: *http://www.springerlink.com/content/937650444v703512/*

Hietanen, O., Lefutso, D., Marais, M., Munga, N., Taute, B., Nyewe, M. & Semwayo, T. (2011). How to create national foresight culture and capacity: case study South Africa, *Ekonomiaz* 76(1): 144–185.
URL: *http://hdl.handle.net/10204/5062*

Hill, D., Doody, L., Watts, M. & Buscher, V. (2011). The smart solution for cities.

Hodgkinson, S. (2011). Is Your City Smart Enough?, *Technical report*, OVUM.

International Telecommunications Union (2005). *ITU Internet Reports 2005: The Internet of Things. Executive Summary*, ITU, Geneva.
URL: *http://www.itu.int/osg/spu/publications/internetofthings/*

IOT-A: Internet of Things Architecture (2011). Accessed December 25, 2011.
URL: *http://www.iot-a.eu*

IOT-I: Internet of Things Initiative (2011). Accessed December 25, 2011.
URL: *http://www.iot-i.eu/*

Macagnano, E. (2008). Intelligent urban environments: towards e-inclusion of the disabled and the aged in the design of a sustainable city of the future, a South African example, *Fifth International Conference on Urban Regeneration and Sustainability: The Sustainable City 2008*, Skiathos, Greece, pp. 537–548.

McCullagh, P. J. & Augusto, J. C. (2011). The Internet of Things: The Potential to Facilitate Health and Wellness, *UPGRADE. The European Journal for the Informatics Professional* XII(1): 59–68.

Nosowitz, D. (2011). Everything You Need to Know About Near Field Communication. Accessed December 25, 2011.
URL: *http://www.popsci.com/gadgets/article/2011-02/near-field-communication-helping-your-smartphone-replace-your-wallet-2010/*

O'Brien, T. (2010). In a Nutshell: What are QR Codes? Accessed December 25, 2011.
URL: *http://www.switched.com/2010/06/21/in-a-nutshell-what-are-qr-codes/*

Robles, R. J. & Kim, T.-h. (2010). Applications, Systems and Methods in Smart Home Technology: A Review, *International Journal of Advanced Science and Technology* 15: 37–48.

Schaffers, H., Komninos, N., Pallot, M., Trousse, B. & Nilsson, Michael Oliveira, A. (2011). Smart Cities and the Future Internet: Towards Cooperation Frameworks for Open Innovation, *in* J. Domingue, A. Galis, A. Gavras, T. Zahariadis, D. Lambert, F. Cleary, P. Daras, S. Krco, H. Müller, M.-S. Li, H. Schaffers, V. Lotz, F. Alvarez, B. Stiller, S. Karnouskos, S. Avessta & M. Nilsson (eds), *The Future Internet*, Vol. 6656 of *Lecture Notes in Computer Science*, Springer Berlin Heidelberg, pp. 431–446.
URL: *http://www.springerlink.com/content/h6v7x10n5w7hkj23/*

Secretariat for the Convention on the Rights of Persons with Disabilities (2011). Convention on the Rights of Persons with Disabilities.
URL: *http://www.un.org/disabilities/default.asp?id=259*

Sense Smart City (2011). Accessed December 25, 2011.
URL: *http://sensesmartcity.org/*

Seventh Framework Programme (2011). European Research Cluster on the Internet of Things. Accessed December 25, 2011.
URL: *http://www.internet-of-things-research.eu/*

SmartSantander (2011). Accessed December 25, 2011.
URL: *http://www.smartsantander.eu/*

Stephanidis, C., Salvendy, G., Akoumianakis, D., Arnold, A., Bevan, N., Dardailler, D., Emiliani, P. L., Iakovidis, I., Jenkins, P., Karshmer, A. I., Korn, P., Marcus, A., Murphy, H. J., Oppermann, C., Stary, C., Tamura, H., Tscheligi, M., Ueda, H., Weber, G. & Ziegler, J. (1999). Toward an Information Society for All: HCI Challenges and R&D Recommendations., *Int. J. Hum. Comput. Interaction* pp. 1–28.

Sundmaeker, H., Guillemin, P., Friess, P. & Woelfflé, S. (2010). *Vision and Challenges for Realising the Internet of Things*, European Commission.
URL: *http://www.internet-of-things-research.eu/pdf/IoT_Clusterbook_March_2010.pdf*

Taylor, A. S., Harper, R., Swan, L., Izadi, S., Sellen, A. & Perry, M. (2007). Homes that make us smart, *Personal Ubiquitous Comput.* 11: 383–393.
URL: *http://dx.doi.org/10.1007/s00779-006-0076-5*

Vermesan, O., Friess, P., Guillemin, P., Gusmeroli, S., Sundmaeker, H., Bassi, A., Jubert, I. S., Mazura, M., Harrison, M., Eisenhauer, M. & Doody, P. (2011). *Internet of Things – Global Technological and Societal Trends*, River Publishers, chapter Internet of Things Strategic Research Map. ISBN: 978-87-92329-67-7.

Web Accessibility Initiative Guidelines and Techniques (2011). Accessed December 25, 2011.
URL: *http://www.w3.org/WAI/guid-tech.html*

Weiser, M. (1991). The Computer for the Twenty-First Century, *Scientific American* 265(3): 94–104.

Williams, Q., Greef, M., Boshemane, S. & Alberts, R. (2008). Designing future technologies for disabled people in a developing country, *in* P. Cunningham and M. Cunningham (ed.), *IST-Africa 2008 Conference Proceedings*, Windhoek, Namibia.

4

An Implementation of Intelligent HTMLtoVoiceXML Conversion Agent for Text Disabilities

Young Gun Jang
Chongju University
Republic of Korea

1. Introduction

HTML represents information in a visual manner and as a result text disabilities such as visually impaired individuals and dyslexics are unable to access their content. Beyond doubt the vast majority of modern web applications neglect the special needs of disabled people. The acoustic representation of the web content contributes to the purpose of pervasive learning and offers great help to disabled individuals. Proposed application of this paper is based on VoiceXML in order to become accessible acoustically by a standard phone or by a computer.

Almost all internet contents are constructed by HTML and the contents which internet user prefer are various and extensive. Manual contents conversion from HTML to VoiceXML for the general public needs so much costs and seems to be practically impossible. Therefore automatic or semiautomatic conversion is very necessary for practical acoustic service.

The possible applications on the internet and intranet are as various as the contents contained on the internet and as the users who utilize the information. This variety increases the attempts of making an automated internet for its complexity of information and service. Especially, these attempts are very necessary for the web contents processor to decrease the cost of producing and managing web contents which is increasing rapidly, also to minimize the number of mistakes caused by human error. Including filling out forms, web automation should automatically operate the web activities, and it should also operate according to the plans or requirements. However, information on the internet often changes its contents, and even its structure. Because HTML describes more for visual expression than for the contents or structure, web pages are still unable to be interpreted perfectly by present technologies (Asakawa, 2000).

VoiceXML is an eXtensible Markup Language (XML) based language that aims to function as a tool for the development of interactive voice response applications. VoiceXML is designed for creating voice applications that feature synthesized speech, audio recognition of voice input or Dual-Tone Multi-Frequency input, recording of spoken input, reproduction of audio files, control of dialog flow, and telephony features, such as call transfer. In general terms VoiceXML attempts to constitute the audio equivalent of

Extensible Hypertext Markup Language (XHTML) for the voice web instead of the visual web.

Before VoiceXML, VoxML which is one of the original types was published by Motorola, also Goose, et al was the first to convert HTML into VoxML. Goose, et al presented the study of Vox portal which makes it possible to access the web through a phone-line, also has 4 layer structures such as client, agent, web server, database, which was added to the VoxML-Agent that can convert HTML to VoxML in the traditional 3 layer WWW structure. Concerning the VoxML-Agent, the interaction with a Vox portal was mainly mentioned, however, the core part which is the design of the conversion function was not mentioned (Goose et al., 2000).

Building a vocal interface by interpreting HTML documents, which use mainly visual interface, is currently being studied (Mohan et al, 1999; Asakawa et al., 1998; Vankayala et al., 2006), also code conversion based on the grammar characteristics is being studied mainly with a tool to adjust and access the internet contents (Embley et al., 1999), code conversion based on the structure characteristics of HTML (Goble et al., 2001; Huang, 2000; Buttler, 2001; Choi, 2001), Semantic code conversion(Hori et al., 2000; Krulwich, 1997), manually adding annotation code (Asakawa, 2000; Lieberman et al., 1999), code conversion used structure and format information (W3C, 2000), semi automatic code conversion using the interaction between structure model and computer (Li et al., 2004), however, these studies are not being generalized because the purpose of the study is to rearrange web pages and convert web objects for the easier voyage and understanding and only to solve a specific user group's needs. It is impossible for not only computer but also humans who know HTML well to guess visual layouts, and to separate each content perfectly just by reading HTML code. Therefore, it is necessary to limit the converting subject and intelligent methods of selecting, extracting and separating the subjects. The converter selects the contents, separates and extracts selected contents, and converts the extracted contents into a VoiceXML document according to the priory defined scenario. The core technology of the converter is to select, and extract the contents that the users want, regardless of the expression of HTML and to accurately separate the contents according to the statistical, grammatical, and structural characteristics of an HTML document and to increase the automatic rate of converting contents by one time access for the practical purpose. Regarding the separation of contents, Embley (1999,2006), Buttler (2001) and Choi (2001) suggested a method of separating multi contents that are arranged in a row with the similar contents like a bulletin board, according to the contents in the HTML document, that has multi contents, and also suggested Heuristic algorithms that can extract separation tags which is the standards of the separation by using the structural characteristics of an HTML document. For the limitation of the suggested Heuristic algorithm, Choi (2001) limited the subjects of the multi contents type to bulletin boards, lists and searching results. The limited subjects should be selected safely on the web page for the successful conversion. However, in the case of using the structural characteristics to select the contents, the contents can be selected differently according to the various expressions and changes of contents on a web page. The minimum child node that is used by Embly and Goose can't properly reflect the user's intention, also the minimum child node that is suggested by Choi and node selection methods which combines included characters and numbers can make different selections according to the number of nodes and the balance of included numbers of characters. Therefore, other methods are required to use the structural characteristics. In the semantic approach, it suggests using ontology for this problem (Huang, 2000; Goble et al., 2001). The documents should be described on an XML base because it is based on the meaning, also it requires an RDF to interpret the ontology. However, it is very complicated to describe the ontology to select contents on the web page that we already know, and it requires another RDF and professional knowledge about the technical language and grammar, moreover, if it doesn't exactly match with the ontological specifications, it responds to the slight changes of the contents very sensitively (Goble et al., 2001). To reflect the user's intention, Krulwich (1997) used the method in which the agent records the interaction between the user and the web page so that, when the user connects to the web page again, the agent performs the same work that is selected by the user. However, this method is very difficult to implement, also it uses the HTML location defining language as the converted result during the automatic process. Lieberman of MIT suggested a way to make the agent recognize the characters through training (Lieberman et al., 1999). This method removed the professionalism that is necessary while manually entering the grammar and eliminating the possibility for mistakes by suggesting examples to the agent and by training the semantic character patterns which are in the unstructured web information. These studies concerning the selection of contents all have some advantages, however, they are not very convenient and strong because they might require unnecessary information to be defined by the users or provide unnecessary interface, also they are very complicated to implement.

This study suggests; 1) the hybrid sequential contents group selection method that utilizes a variety of factors including structural characteristics, the users' prior knowledge, interaction with the agent, and character information to ensure the inclusiveness and definiteness of the contents based access methods to solve the problem of selecting contents accurately, and it also maintains advantages such as there is no need for professionalism and convenient character recognition and it covers the disadvantages. 2) the sequential recognition for documents based on schematics that provide documents interactively according to the user's prior knowledge regarding the relationship between the documents, which is to increase automation rate of the converter. 3) converting the entire structure of HTML to VoiceXML, which is applied the above suggested 1) and 2), according to the content separation which is the basic system and designated scenario. In Section 2, it is suggested the document type for conversion, in Section 3 and 4, it is suggested the contents selection method and recognition methods of sequential web documents and in Section 5, it is suggested the converting from HTML to VoiceXML agent structure which is combined the suggested methods, and constructing methods, also described the contents separation method.

2. Selection by the convertible HTML document type and structural characteristics

A web author tries to put a lot of information on one page by using various visual effects. This information is visually categorized and then is fragmented, For example, a web page in Fig. 1 has A, B, C, D, E, F, G and H fragments on one page. It is necessary to determine which fragment is the main information on this page and to separate the information according to the contents and then it should be converted to VoiceXML document (W3C, 2000).

In order to convert HTML to VoiceXML, first of all, the HTML contents should be separated into fragment units according to the contents, then a VoiceXML document should be created for each fragment. However, HTML tags are difficult to separate into fragments because they only express how to visualize the information, and don't include any meaning as an XML tag.

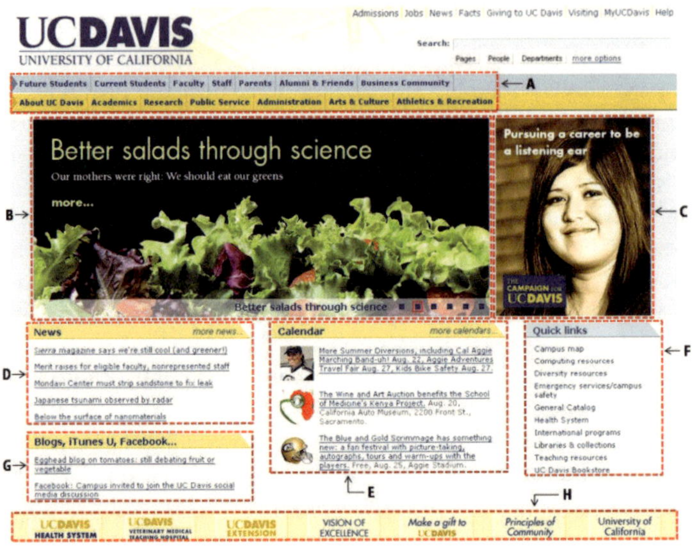

Fig. 1. Web page type.

Therefore, it is assumed that the contents are visually categorized. The probability of this assumption was proved through the code verification of most web pages. For the detailed separation of the contents into fragments, the heuristic method was used, which statistically analyzed the structural characteristic of the tags. Therefore, the types that can be separated by using this method are limited, also the qualitatively similar types of contents are arranged in a row. The documents types that can be applied the conversion method of this study are bulletin board, list and search result. This type of HTML document has many features that are similar to child nodes when the documents are expressed in a tree structure, also the contents include many characters, which means the location of the contents can be found. The contents that can be converted to VoiceXML in Fig. 1 are the News(D), Calendar(E) and Blogs, ITunes U, Facebook(G).

To analyze the structure of an HTML document, for the first step, it must be reorganized into a tree structure. The reason is because it is easier to analyze an HTML document in a tree structure.

The minimum reformed sub tree method is designed as the selection method of the main content group by using the structural characteristics on the web page. The minimum sub tree means the smallest tree that includes contents of the whole tree structure. Before the contents are extracted, it first extracts the sub tree that includes the contents, and this sub tree corresponds to each fragment of Figure 1. The minimum sub tree is the tree that uses the nodes with the most child nodes as the path after finds out how many child nodes are in each node. Usually, in the web page where contents are arranged in a row, the possibility that the main contents are included in the sub tree that uses the node with the most child nodes as its path, is very high. However, if the numbers of child nodes becomes the main concern of extracting the minimum sub tree, in the web page which has lots of menus, the minimum sub tree can be the sub tree that includes the menu.

In this paper, it is suggested that the text size of the node with the most child node be used to extract the minimum sub tree. Meaning that first it is necessary be find the node with the most child nodes, then figure out the text size of each node and finally the minimum sub tree should be the sub tree that uses the node with the biggest text size as its path. In this method, we can avoid the mistake of designating the sub tree that has most child nodes with the small text size, like a menu, as its minimum sub tree.

3. Selecting an interactive contents group based on contents

When selecting a contents group through the structural interpretation of an HTML document, it is difficult to apply the user's intention with any kind of rule platform, and difficult to select multiple contents on one web page. Therefore, there must be a mechanism to select contents again by reflecting user's intention if the automatically selected contents by using the structural characteristics are not reflecting user's intention. The suggested method is the sequential selection method which provides the automatic selection and the user's multi selection according the frequency of use based on the interactive selection with users. It first applies the structural method to select the proper item, and if it is improper, then it can be selected suggesting examples through the character information and by checking the composed results on the web page.

At this time, the suggested character row becomes the title of the documents that are created as the result of the conversion. Also if the wrong contents group is selected, it can be selected by designating numbers or numbers with logical calculus. In this paper, character rows included in a web page were used to train the agent. Web authors usually try to show the contents to users in a simple and effective way by using key words as characters or graphics. Therefore, in this method, it uses the author's expression ability and insight about the web page that is showed to the users, and copy key words are selected for the contents to transfer to an agent, also it designates a sub tree that includes the valid key words among the already written sub tree as the converting subject for the structural interpretation. However, it is unable to copy the characters of key words if they are expressed graphically, and this method can't be used if multiple contents or logically composed contents need to be selected. Contents composition looks like the wrong selection or a visually same contents group, however, it can be used very effectively on web documents that are made with a different structure. If there is a problem for selection from the character row, it can include the index number of contents, which are made according to the result of structure interpretation, on the original document, and can visually provide the composed web document to the user, also it can select the contents by transferring the index number to the agent according to user's judgments. Multiple selections is possible in this method, also it supports logical composition + operator which means combining is only used for the logical composition operator, however, it can be added by needs.

Fig. 2 is an example of the valid implementation algorithm. In this method, the structure interpretation method as tree structure combined with the rule platform that uses the amounts of nodes and characters, also if the automatic selection is failed because it doesn't reflect user's intention: there are many web making methods that can have a similar visual effect: it is impossible to use for all methods that can be used on HTML interpretation, so it utilizes a hybrid sequential method with a meaning based access that allows users to interactively select the interpreted results of the system.

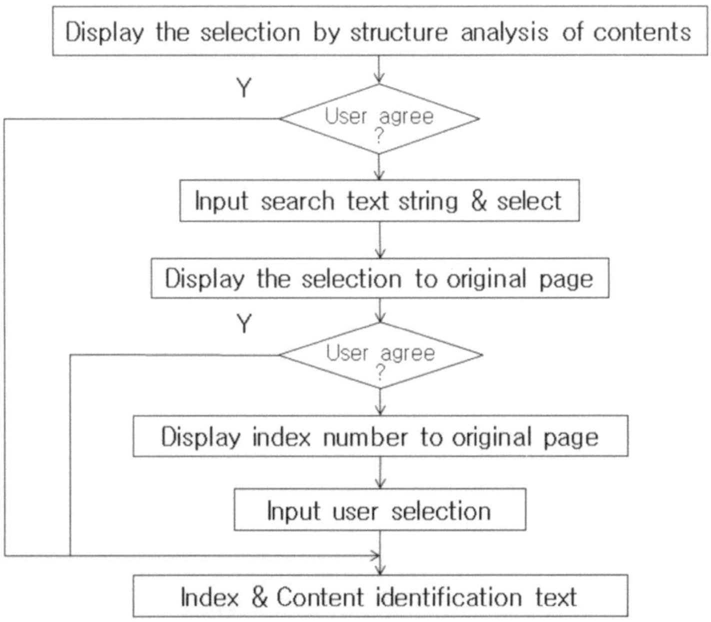

Fig. 2. Flow of the hybrid sequential contents group selection method.

4. Sequential recognition of web document based on interactive schema utilizing the prior knowledge

Generally, if there are many contents that have the same characteristics and structure, the contents should be made with several web documents and those web documents should be connected through the link because the amount of contents that can be expressed on one web page is limited. However, it is unable to determine which link is about the next document by only looking at the HTML code, so it is impossible to extract automatically at once. To solve the above problem, the entire contents can be extracted by interactively providing the user's prior knowledge regarding the link between the documents, and this information doesn't designate the tag path by analyzing the HTML code, but provides the link information like the link is located after how many links from the last contents so the URL for the next document can be extracted. At this time, the link between documents is usually located after the last contents, and the link that connects the next document can be divided into three categories; text, image and serial number, as seen in Fig 3.

The prior knowledge information input scheme can be divided into 3 steps as seen in Fig. 4. In the first step, the link type for the next document is designated, and in the second step, the details such as link character, link location and serial number orders is designated according to the link type, and in the last step, the location of the last document is designated, like how many documents will be extracted.

Fig. 3. Linking type for the connected page and link location.

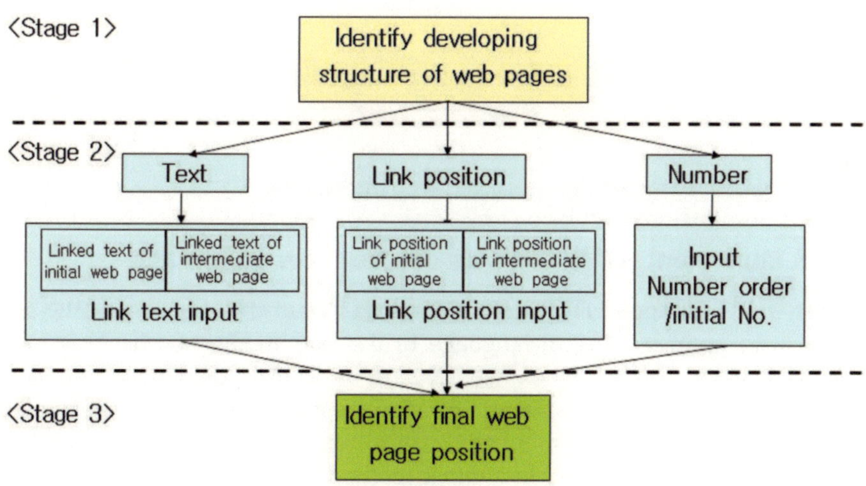

Fig. 4. Prior Knowledge Information Input Schema.

When the prior knowledge is provided, the URL for the next linked document can be extracted through the procedure in Fig. 5.

The prior knowledge information input procedure for the HTML document structure is provided in a Windows Wizard style. The Prior knowledge information input wizard can be divided into 3 steps as the Prior knowledge information input schema. In the 1st step, the link type information should be entered, then, in the 2nd step, if the link type is text, enter the text information, if it is image, enter link location information, and if it is serial number, enter serial information. In the 3rd step, enter the final web document information which is needed to be extracted.

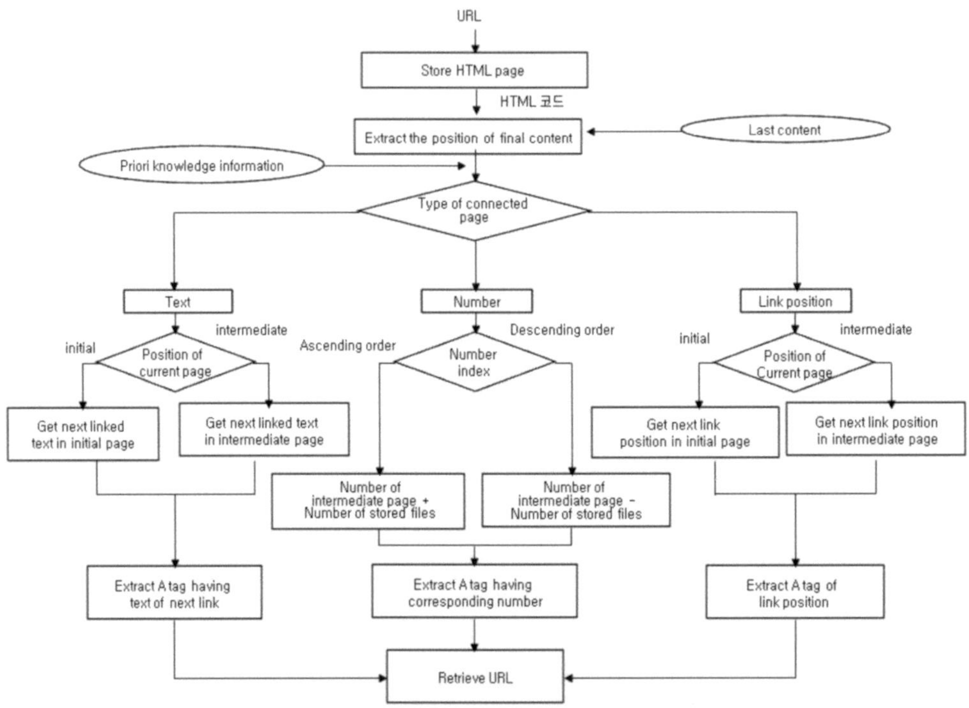

Fig. 5. Extracting URL Flow for the connected web documents.

5. Design & implement HTML to VoiceXML conversion module

The simplest way to change HTML into a markup document is to match the tags with similar roles with another markup language to the tags in the HTML. This method of conversion is usually applied when we convert an HTML document into a markup document for wireless internet. However, HTML tags don't provide any meaning but the visual expression of the contents, so it is difficult to convert and provide exact information by using 1:1 matching for each tag. In another method, we can reorganize the scenario by recording the web documents path and find the contents using that path so the information can be provided with the markup language. At this time, the web document path will be recorded as the absolute path that uses html or a body tag as its path in the tree structured HTML(Freire et al, 2001). The problem of this method is that it can cause unexpected results because the tags were deleted, modified, and inserted for the path. Embley, Butler and Choi suggested separating the multiple contents that are arranged into similar types such as bulletin board or searched results by using the documental structure, however, it is only about one web document, so we need to connect each web document to separate the sequential contents across several web documents.

The HTML to VoiceXML Conversion Agent that is implemented in this paper can be divided into three procedures as seen in Fig. 6; Extracting list and creating VoiceXML

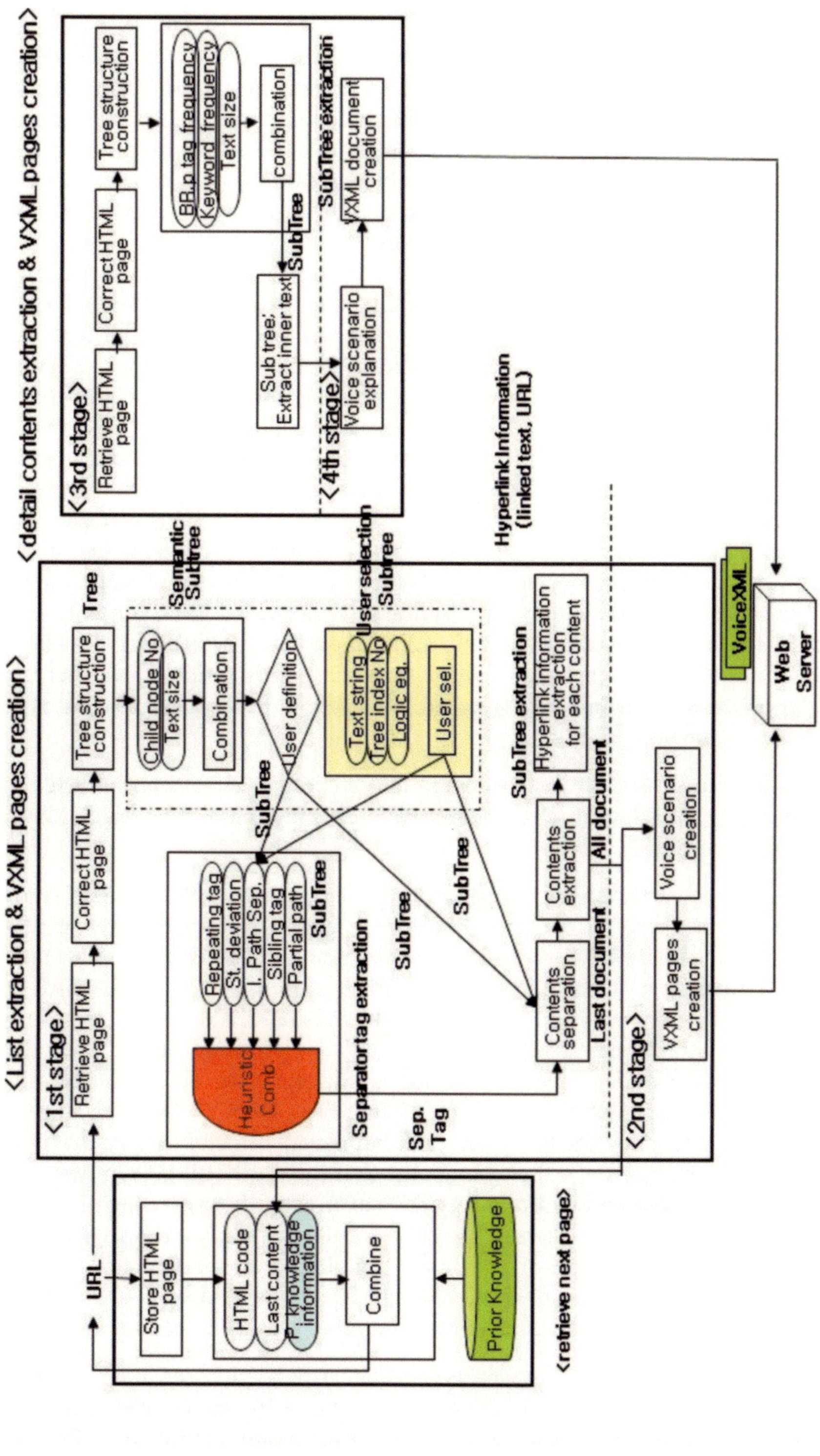

Fig. 6. HTMLtoVoiceXML Conversion Module Map.

document, Extracting details and creating VoiceXML document, and Extracting URL of the next connected web document. Extracting list and creating VoiceXML document procedure is extracting list contents from the list HTML document and then creating the VoiceXML document that is valid for the list scenario. The dotted line part is selected contents after the conversion in the tree structure in the first step, and it is implemented through the hybrid sequential contents selection algorithm that is suggested in this paper. In this step, the contents selection method which has faster processing speed and suggested in section 2, will be applied, also if it is not the contents wanted, it will be applied the method that is suggested in section 4. In the process of extracting details and creating a VoiceXML document, it extracts the details from the detail contents document about the list contents and creates it with VoiceXML documents which will be valid for the detailed contents scenario. In the process of extracting the URL from the next connected web document, if the contents are across several web documents, it extracts the list and creates the VoiceXML document after finding out the URL of the connected web page. It finds the URL of the connected web document through prior knowledge of the list HTML document structure, extracts the list, and creates the VoiceXML document to extract all the contents during a single connecting by transferring the value. This part has not been suggested in other existing studies, however, it extracts connected web documents by using the connection rule which determined by the connected web document recognition method, then transferring that information to the 1st step and continuing to perform the same procedure. The dotted line part which is the content selection block and the solid line part which is extracting URL block for the connected web page in Fig. 6 opposite with the data block that uses the fist interpretation result during the repeated connecting procedure, also it is connected continuously by the scheduler.

First of all, the candidate tags for the separation of contents should be extracted in order to separate contents to each unit. The contents candidate tags should be the child nodes of the path nodes of the selected sub tree.

If there is only one candidate tag for separating the contents, that candidate tag will be the boundary tag to separate the contents, however, if there are more than two candidate tags, then the valid tag should be selected as the separation tag. To extract a valid separation tag from several separation candidate tags, the heuristic algorithm was used, which is based on a few rules used when making HTML document. The heuristic that is used in this paper makes: standard deviations about the amounts of characters between candidate tags: repeating patterns of the tag pair which consistently shows on each contents: all path lists from candidate tag nodes to other temporary nodes then, counts how many times the partial paths that are appeared by the identified path happened and sibling tags that adjoin to sub tree happened, then, rearrange the sibling tags that categorize all tags in descending order and tags which are frequently used to separate contents in the ascending order, then finally use the identifiable path separator tag(IPS) for the candidate tags. Each heuristic is optimized for the special type of web page, but it is independent from other heuristic algorithms. Therefore, it is considered to combine each independent heuristic to increase the possibilities of extracting correct contents separation tags of web document.

To decide on the best combination status for 5 heuristic algorithms, the Stanford Certainty Theory was used (Luger et al., 1997). There are a total of 16 ways to combine each heuristic algorithm, however, most the exact way was to extract the separation tags after combining 5 heuristic algorithms.

When the contents are separated for each separation tag, the starting and ending points of the contents should be extracted. Meaning that it is necessary to first decide if the starting tag of the separation tag and the next starting tag of the separation tag should be the starting and ending point, or if the contents right before the separation tag and the starting tag of the right next separation tag should be the starting and ending point to separate contents. Separating procedure with the standard of the separator tag, which is suggested in this paper, is as following.

```
1. Find the first child node N of the path node of the minimum sub tree
2. repeat
if (N == STYLE Tag or SCRIPT Tag or !(Annotation) Tag)
or ( N == BR Tag and BR Tag != Separation Tag)
or ( N == HR Tag and HR Tag != Separation Tag)
or ( N == P Tag and P Tag != Separation Tag)
then
N := Next siblingnode
continue /* Perform the if sentence again */
else
End the repeat sentence
end if
until end-of-Minimum sub tree
3. if N == Separation tag then
Contents := The contents between the starting tag of the first separation tag
and the starting tag of the next separation tag
else
Contents := The contents between the starting tag of the first separation tag
and the starting tag of the next separation tag
end if
```

When the contents are separated with the standard of the separator tag, the invalid contents can be included. Therefore, the valid contents should be extracted from the separated contents. In each text of contents of the web page, there is an arrangement of regular types of numbers, dates and characters, so it is necessary to use these characteristics to extract the valid contents. The suggested method of extracting valid contents in this paper is as follows:

```
1. repeat /* For all separated contents */
Text type of the separated contents should be divided into numbers, dates
and characters to be saved at F arrangement
until end-of-Separated contents
2. Find the amounts of the same types from the saved information at F
arrangement.
3. Find the most types of information and extract the contents that have this
type of information.
```

The extracted contents through the extracting procedure shall be provided to the users through the vocal interface. Because users can understand only a limited amount of vocal information at once, compared to visual information, N vocal scenarios shall be created according to the amount of extracted contents.

In this paper, if the average characters of the extracted contents is less than 100, a vocal scenario with 4 contents will be created, and if it is more than 100, N scenarios will be created consisting of 2 contents in one vocal scenario.

Each vocal scenario creates VoiceXML document according to the VoiceXML 1.0 format. The created VoiceXML document has the basic structure displayed in Fig. 7.

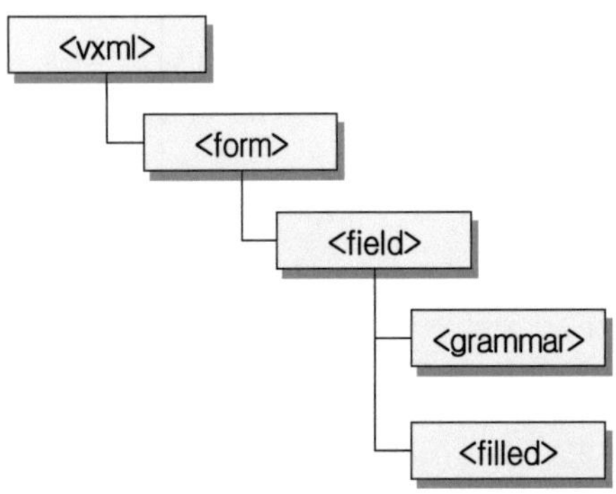


```
<?xml version="1.0"?>
<vxml version="1.0">
<form>
<field>
List contents prompt
Menu prompt
<grammar>Vocal menu item</grammar>
<filled>
'If' sentence according to the menu
selection
</filled>
</field>
</form>
</vxml>
```

Fig. 7. Basic structure of the created VoiceXML document.

6. Test & results

To test the performance of the HTML to VoiceXML conversion agent that was implemented in this paper, board type, list type and search result type employment sites, newspaper sites and search engines with high access frequency were selected. The specific sites selected for test are listed in table 1, and the sites that were tested for all 3 types are listed in table 2.

The test method is the same as the method in GIT, and the succeed rate of algorithm will be counted in two steps (Li et al., 2004). First, the suggested algorithm was applied to each web site page, the tag that with the highest score was found and the exact rate of the tag on that page was counted. Second, an average of all web sites was calculated to see the success rate of the heuristic algorithm and the combined algorithm.

In the result of applying this method to 200 web pages, which are shown on table 1, the separation successful rate was 100% except for the pages that described the wrong HTML grammar. This was because most of the web sites consisted of tables, and in many case, it had only one separation candidate tag, so the heuristic algorithm was not needed.

Division	Name of Web site	Web site URL
Employment (Board type)	Incruit	www.incruit.com
	Recruit	www.recruit.co.kr
	Hellojob	www.hellojob.com
	Jobex	www.jobex.co.kr
	Devpia	www.devpia.com
Daily News (List type)	DongA Daily	www.donga.com
	JoongAng Daily	www.joins.com
	Sports Today	www.sportstoday.co.kr
	GoodNews	www.kukminilbo.co.kr
	Hankyoreh	www.hani.co.kr
Search (Search results type)	MSDN	msdn.microsoft.com
	Paran	www.paran.com
	Altavista	www.altavista.co.kr
	Google	www.google.com
	Naver	www.naver.com

Table 1. Test subject web sites.

Name of Web site	Web site URL
Hankooki	www.hankooki.com
Hankook Ilbo	www.hankooki.com/hankook.htm
Daily Sports	www.hankooki.com/dailysports.htm
Seoul Economy	www.hankooki.com/sed.htm
Korea Times	www.hankooki.com/sed.htm

Table 2. Test subject web sites for success rate.

Heuristic Algorithm	No. 1	2	3	4
Separative Tag	0.85	0.1	0.15	0.0
Repeating Pattern	0.65	0.0	0.15	0.0
Partial Tag	0.95	0.0	0.0	0.0
siblingTag	0.80	0.2	0.0	0.0
Standard Deviation	0.70	0.2	0.1	0.0
Combination Algorithm	0.99	0.02	0.01	0.0

Table 3. The probability ranking of each algorithm and the success rate of the combined algorithm.

Therefore, we tested the success of the separation tag extraction rate for the web sites that used various web page writing methods. In table 3, the probability ranking of each heuristic algorithm and the success rate of the combined heuristic algorithm is described for about 200 web pages from the web sites suggested in table 2. The reason it had a higher succeed rate compared to the GIT combination algorithm success rate of 94%, is because the different types of web page were less than GIT. It is difficult to compared with the GIT case, because there is no information about the web page. However, the test results show that the implemented converter is very trustworthy for the regular table or the contents that links are arranged in a row, also when the combined heuristic algorithm was applied for the web page that used various methods, it showed a 99% success rate, so most of the web pages were convertible. In table 3, it shows the success rate when each heuristic algorithm is applied for the separation candidate tags of the web pages, and when each heuristic is combined.

Web Site Name	Prior Knowledge Type
Incruit	Numbers, Image
Recruit	Numbers, Image
Helljob	Numbers, Image
Jobex	Numbers, Text
Devpia	Numbers
DongA Daily	Image
JoongAng Daily	Image
GoodNews	Image
Hankyereh	Numbers
MSDN	Numbers
Paran	Numbers
Altavista	Numbers
Google	Numbers, Text
Naver	Numbers

Table 4. Prior Knowledge type of each Web Site.

The test regarding extracting web documents by using the prior knowledge was performed for the web sites in table 1. As we can see from Fig. 5, the most popular type among the types of prior knowledge for the connected web document was serial number type, and the next most popular types was mixed types like serial numbers and image. Also the success rate for using prior knowledge to extract web documents was 100%.

7. Conclusion

This chapter is regarding a conversion system using VoiceXML to provide the HTML contents vocally through a mobile terminal or phone for text disabilities. To decrease the cost and time needed to convert to VoiceXML, the HTML to VoiceXML conversion agent was designed and implemented that is capable of automatically converting HTML documents into VoiceXML documents.

Because it is unable to figure out the meaning of the information on HTML documents, the convertible HTML document type is provided and a hybrid sequential contents selection method was suggested to solve the problem of the stiffness of using structural characteristics and that fail to reflect the user's intention and its effectiveness was proved. Also the contents were separated and extracted to be converted into VoiceXML documents according to the vocal scenario by analyzing the document structure. In addition, the same type of contents across several web documents was extracted at once through the prior knowledge regarding the web linking structure to create practical and effective vocal scenario.

The function and performance of the developed converter were tested on about 400 Korean web pages. In the results, all of the web pages that applied HTML grammar correctly were able to be converted into VoiceXML documents. Therefore, it is determined the application is effective, practical and success rate.

A few web pages that didn't follow exact HTML grammar, especially, if there was no ending tag for the starting tag, were not converted correctly, and to solve this problem, the HTML should be converted into XHTML documents that kept the XML format, and then converted into VoiceXML documents. The presently implemented converter is limited to the convertible HTML documents type because of the inherent problems, so a more intelligential method should be combined to expand the subjects of the conversion.

8. References

Asakawa, et al, (1998). User Interface of a Homepage Reader, Pro. of ASSET'98, pp149-156.

Asakawa (2000). Annotation-Based Transcoding for Nonvisual Web Access, Pro. of ASSETS'00, pp172-179

Buttler, D. ; Liu, L. & Pu C. (2001). A Fully Automated Extraction System for the World Wide Web", IEEE ICDCS-21, pp16-19

Choi, H.I. & Jang, Y.G. (2001). Design and Implementation of a HTMLtoVoiceXML Converter" , Journal of KISS : Computing Practices, Vol. 7, No. 6, pp559-568.

Embley D.W.; Jiang Y.S. & Ng Y.K. (1999). Record- boundary discovery in Web documents, Proceedings of the 1999 ACM SIGMOD International Conference on Management of Data , pp467-478

Embley, D. W. et al, (2006). Notes contemporary table recognition", DAS 2006. LNCS 3872, pp. 164-175

Goble, C. & Bechhofer, S. et al, (2001). Conceptual Open Hypermedia = The Semantic Web?, Proceedings of the 2nd Int. Workshop on the Semantic Web

Goose, S.; Newman, M.; Schmidt, C. & Hue, L. (2000). Enhancing Web accessibility via the Vox Portal and a Web-hosted dynamic HTML<->VoxML converter, WWW9, Vol. 33, No. 1-6, pp583-592.

Hori M.; Kondoh G.; Ono K., Hirose S. & Singhal S. (2000). Annotation-based web content Transcoding, Proc. of WWW9, pp. 197-211

Huang, A. W. (2000). A Semantic Transcoding System to Adapt Web Services for Users with Disabilities", Pro. of ASSETS'00, pp156-163

Krulwich, B. (1997). Automating the internet: agents as user surrogates, IEEE Internet Computing, Vol. 1, No. 4, pp. 34-38

Li, S. et al, (2004). Automatic HTML to XML conversion, WAIM 2004, LNCS 3129, pp. 714-719

Lieberman, H.; Nardi, B. A. & Wright D. (1999). Training Agents to Recognize Text by Example, Proceedings of Agents'99, pp1-5

Mohan, R.; Smith, J. & Li, C.-S. (1999). Adapting multimedia internet content for universal access, IEEE Transactions on Multimedia, Vol. 1, No. 1, pp104-114.

Vankayala, R. R. & Shi, H. (2006). Dynamic voice user interface using VoiceXML and Active Server Pages, APWeb 2006, LNCS 3481, pp. 1181-1184

W3C (2000). Voice eXtensible Markup Language(Voice XML) version 1.0, http://www.w3.org/TR/voicexml, W3C Note 05

Freire J.; Kumar B. & Lieuwen D. (2001). WebViews: Accessing Personalized Web Content and Services, WWW10, pp576-586

Luger, G.F. & Stubblefield, W.A. (1997). Artificial Intelligence: Structures and Strategies for Complex Problem Solving, Third Edition. Addison Wesley Longman, Inc.

5

Touch Screens for the Older User

Niamh Caprani, Noel E. O'Connor and Cathal Gurrin
CLARITY: Centre for Sensor Web Technologies
Ireland

1. Introduction

"I suspect we will be seeing touch screens used for more applications than ever before" (Shneiderman, 1991).

It has been 20 years since Ben Schneiderman predicted that there would be an increase in the use of touch screen applications yet it has been only in recent years that this prediction has come to pass. The concept of a touch screen computer was first introduced in 1965 by E.A. Johnson who described the possibilities of touch screen technology to support air traffic controllers. In this article Johnson describes how "the touch display coupled to a computer can be considered as a keyboard", a novel approach at the time. Touch screens were brought into the public domain in 1971 by Elographics, Inc. which instigated the development of public touch screen technology such as automated teller machines (ATMs) and information kiosks (Brown et al., 2011). Another milestone in the history of touch technology was the introduction of the personal touch screen computer, HP-150, developed by Hewlett-Packard in 1983 (Sukumar, 1984). The purpose of this early design was to offer individuals an intuitive technology option. Although touch screen systems have maintained this intuitiveness and ease of use over the years, problems that existed with early systems still provide challenges for designers and developers to this day.

It is largely due to the release of the Apple iPhone and iPad that was quickly followed by similar offerings from competitors that we have recently seen a touch screen revolution. Today, touch screen devices are easily available, portable and relatively inexpensive. Furthermore, commercial touch screen technology offers a balance between accessible interaction and aesthetics. This is welcome news for the older user. Although older adults for the most part have positive opinions about technology (Mitzner et al., 2010), they are less likely to use technology compared to younger individuals (Fisk et al., 2009). However, the main predicting variables of technology acceptance for older adults are usefulness and ease of use (Bouwhuis, 2003; Selwyn, 2004; Mitzner et al., 2010). These are variables which would benefit all users of technology.

There are essentially three ways in which a person can use a touch screen device. These are: (1) one user interacting with a device (e.g. a mobile phone); (2) multiple users interacting with one device at the same time; and (3) multiple users interacting with one device at different times (e.g. tablet computer). Although it is worth noting the possibility for multi-user interaction with a touch screen tabletop, such as the DiamondTouch or Microsoft Surface, throughout this chapter we will mostly be referring to one individual interacting with a touch screen system.

In this chapter we will discuss the implications of touch screen technology for the older user. Firstly in section 2 we will look at the definition of an "older adult" and outline the general characteristics and capabilities of this age group. In section 3 we will discuss the advantages and disadvantages of touch screens and refer to research aiming to increase the usability of touch screens for older adults. Finally, in section 4 we will provide examples of touch screen applications that have been developed and researched in the public, healthcare, home and social domain.

2. Characteristics of the older user

Older people are living longer and the older population is gradually increasing. This demographic shift has motivated researchers from various disciplines to work out ways to improve the quality of life for older people, alleviating the effects of ageing through medication and assistive technology (Astell et al., 2010). In a study focused on population projections over 27 European countries, Mamolo and Scherbov (2009) showed that there will be an overall population shrinkage with an expansion of the elderly population. One person out of four is projected to be over the age of 65 by the year 2030. Figure 1 taken from Mamolo and Scherbov's report illustrates the growing population of older individuals between the years 2007, 2020 and 2030. These statistics emphasise the need for researchers to come up with innovative approaches to support older adults and to embrace this group as potential technology users.

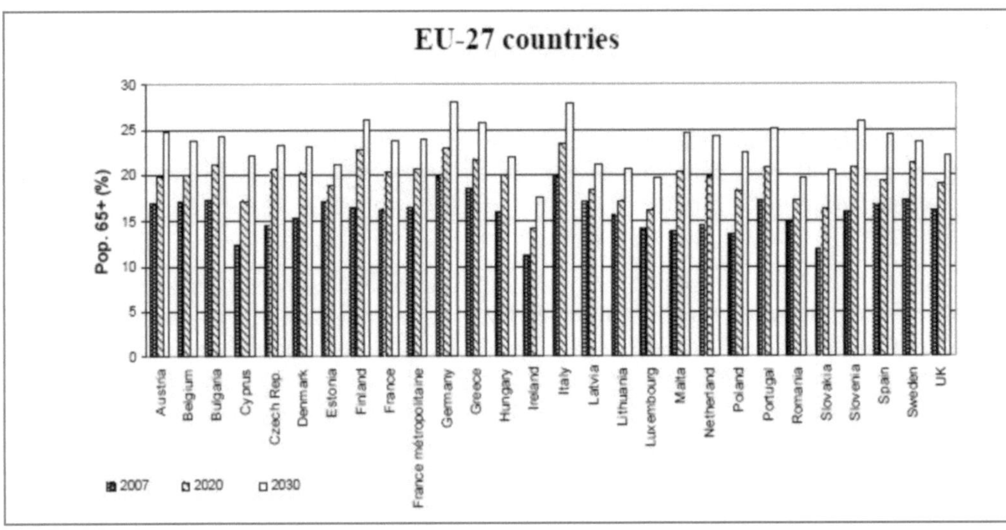

Fig. 1. Population projections for individuals 65+ years for the years 2007, 2020 and 2030, across 27 European countries; Mamolo & Scherbov (2009).

2.1 Defining the "older adult"

Age classification is not a straightforward process as many variables impact the rate at which a person ages. Chronological age markers are the most common measurement used, for example, determining the age of retirement. However this does not take into account biological, psychological and social factors. Newell (2008) groups older people into three broad categories: (1) Fit older people, who do not appear, or consider themselves, disabled but whose functionality, needs and wants are different to those they had when they were younger, (2) Frail older people, who would be considered to have a "disability" and in addition have a general reduction in many other functionalities, and (3) Disabled people who grow older, whose long-term disabilities may have affected the ageing process, and whose ability to function can be critically dependent on the other faculties, which will also be declining. Fisk et al. (2009) state that although individual differences exist, generally older adults have common biological, psychological and social characteristics. They suggest grouping older adults into two groupings: (1) The younger-old, ranging in age from 60-75; and (2) The older-old, for individuals over the age of 75 years. We will briefly outline the characteristics of ageing in the next section.

2.2 Age-related issues and implications for design

Understanding or, at the very least, being aware of the capabilities and limitations of older users can help to guide designers and developers to create more useable technologies. As we grow older we change and develop, increasing some skills, losing others and learning to compensate for those in decline (for review see Bosman & Charness, 1996). Some of these changes have more implications on how we interact with technology (e.g., vision problems) compared to others (e.g., greying hair). The characteristics that we as designers are concerned about include perceptual, psychomotor, cognitive and physical changes (see figure 2 for summary). Perceptual abilities, most commonly for vision and hearing, generally begin to show signs of decline from a relatively young age. Approximately half of all men over the age of 65 and 30% of women suffer hearing loss, and most people notice visual problems around the age of 40 (Fisk et al., 2009). These figures highlight the importance for interfaces to provide clear adjustable output from devices (Charness & Jastrzembski, 2010). Designers of technology systems should accommodate for this by displaying appropriately sized text and design features, using high contrast colours and including adjustable audio output at low frequencies (Fisk et al., 2009; Hawthorn, 2000). Multimodal output would also increase the usability of a design for an older user.

Cognitive changes also have a significant influence for older users. Age-related differences in cognitive functioning can be seen to stem from the reduction of cognitive resources available, impairing older adults' ability to carry out cognitively demanding processes (Kester et al., 2002). For example, working memory, the ability to store and retrieve new information, along with declining information processing speed, effects how older adults learn and interact with new devices. Many systems rely on a person's ability to keep information active however this is unrealistic for older users unless they are proficient users. Therefore designers should make use of appropriate feedback to the user informing them where they are in the system and where they have been. Simple use of text, colour and icons can significantly reduce confusion and increase ease of use. An age-related deterioration of attention skills also means that extraneous or distracting information on screens can cause further complications for the older user.

Older adults show changes in their physical abilities, due to loss of muscle mass and flexibility. Physical problems can also occur as a result of accidents and falls (common with frailer older adults) and age-related conditions such as arthritis and stroke. In computer use this can result in difficulties grasping a mouse, and positioning and controlling the curser

(Hawthorn, 2000). Even with "normal" ageing there is an overall slowness of movement and older users may find it difficult to make precise selections of small interface targets. Frail older users may also have problems pressing buttons on devices such as television remote controls. Matching the input device to the task can go some way to supporting these issues (Charness & Jastrzembski, 2010; Rogers et al., 2005). To reduce these problems, designers should implement large targets for accurate cursor positioning, reduce scrolling when possible and allow for slower response times. Touch screen interfaces are more frequently being used to assist the technology experience of older adults as they require direct input, require large button targets and eliminate the need for multi-components (e.g., a desktop uses a mouse, keyboard and monitor) (Jin et al., 2007).

Touch screen technology can accommodate for some of these age-related limitations. The technology itself requires that the screen size, even on mobile devices, is larger than non-touch devices. In addition, designers of touch screen interfaces incorporate virtual buttons that are large enough for a finger to press accurately. These features mean that items are larger on the screen, making them (a) easier to see and (b) easier to select accurately. Furthermore, the use of virtual buttons on the screen means that older users do not require as much strength to select a target, and they also do not have to divide their attention between the keypad and the screen.

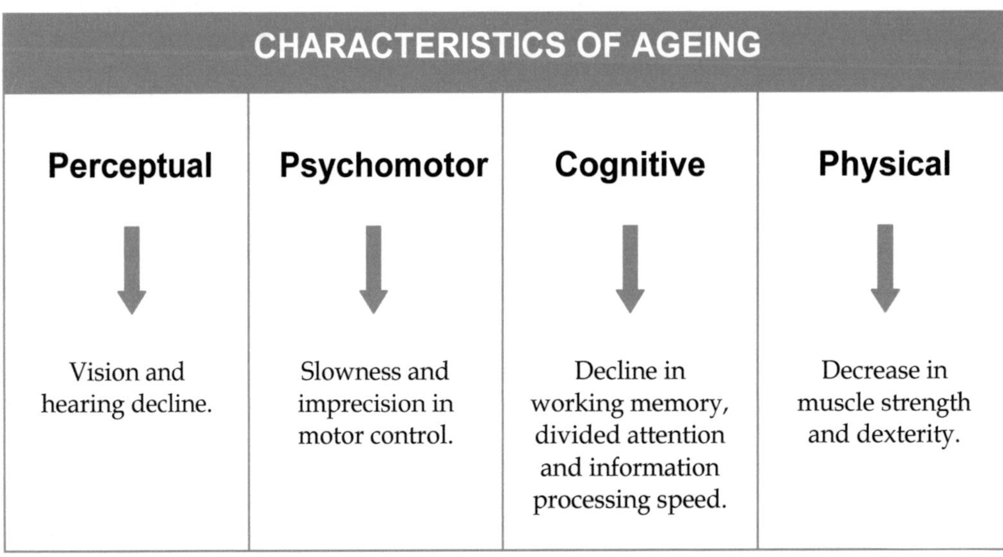

Fig. 2. Summary of age-related changes that have implications on interactions with technology.

2.3 Older adult's use of technology

2.3.1 Technology experience

The number of studies investigating computer use by older adults has progressively increased over the last twenty years (Wagner et al., 2010). This interest stems from a diverse range of research disciplines including, human computer interaction, education and gerontology to name but a few. Although older adults are currently the fastest growing group of Internet users (Wagner et al., 2010), they are less likely to use technology compared to younger groups (Czaja et al., 2006; Czaja & Lee, 2007; Goodman et al., 2003; Morris et al., 2007). Older people generally have a positive attitude towards technology and will use a product if they have a need for it (Fisk et al., 2009). Positive attitudes are also more likely to be expressed towards everyday devices in the home, such as the television, microwave and house alarm (Coleman et al., 2010). Although modern versions of these devices are digital, older users are familiar and comfortable with them. Czaja et al. proposed that the factors predicting older adults' use of computers are age, education, fluid intelligence (abstract problem solving ability), crystallised intelligence (cultural knowledge), computer efficacy (belief about ability), computer anxiety and prior technology experience.

Morris et al. (2007) investigated the older adults' use of computers and the Internet. In total, 473 older adults participated in this survey. The responses showed that word processing and keeping in contact with others (e.g. email) were the most frequently used computer and Internet features. This finding was supported by previous research carried out by Selwyn (2004). Of the Internet users in Morris's survey, 64% stated that it had a positive impact on their lives. The most common reason for not wanting to learn to use a computer or the Internet was that they were simply "not interested". Other reasons given were feeling too old to learn, believing it to be too difficult, and not having access to a computer. Selwyn (2004) found that as circumstances change, so do people's interest or lack of interest in technology. For example, an older person may become interested in using email after their grandchild emigrates abroad, and another older person who used a computer as part of their profession may choose a computer free life after retirement.

2.3.2 Acceptance barriers

So why are older adults less likely to use computers compared to their younger counterparts? As mentioned in the paragraph above, those older individuals who are using computers believe that the technology enhances their lives (Morris et al., 2007) and previous use of technology has an influence on the acceptance of new technologies (Czaja et al., 2006). Similarly, Jay and Willis (1992) tested the attitudes of 101 older adults before and after a two week computer training program. It was found that direct computer experience can significantly modify attitudes, particularly in relation to computer efficacy and comfort. However in Morris et al.'s study, when non-Internet users were asked if there were any factors that would encourage them to learn in the future, the majority (60%) said there was nothing that would influence them. As well as attitude and age-related changes to perceptual, cognitive, and physical abilities acting as barriers for older adults learning new technologies, there are the issues older people have in relation to their privacy, particularly regarding monitoring technologies (Charness & Boot, 2009).

In some situations, older adults are forced to interact with technology to avail of services. For example, many transport services require travellers to purchase their ticket from a machine. Some airlines even have a fee to check-in at person serviced desks to encourage customers to check-in online. Instead of using these technologies it is more than likely that older non-computer users would either avoid the situation or ask a friend or family member to do it for them (Coleman et al., 2010). If users are not given a choice of transaction method, then it is crucial for the available systems be user friendly, intuitive and accommodate for disabilities. In the next section we will outline the advantages and disadvantages of using touch as an input strategy and look at design guidelines for older adult interaction.

3. Touch screens versus other pointing devices

3.1 Advantages and disadvantages

One of the most commonly used approaches for evaluating the benefits of touch screen technology is to compare it to existing methods. As briefly mentioned in the previous section, the suitability of the device may depend on the task at hand. For example, direct input devices such as touch screen may be suitable for novice users for point-and-click tasks, and experienced users may find using indirect devices such as a mouse more suitable for tasks requiring extensive keyboard entry (Charness & Jastrzembski, 2010; Fisk et al., 2009; Rogers et al., 2005). In a study examining the effect of direct and indirect input devices on attentional demand, McLaughlin et al. (2009) found that mismatching the input device to task requires high attentional demand of the users. McLaughlin et al. found that matching the appropriate input device to a particular task (e.g. touch screen for pointing tasks) has considerable benefit for older adults, who experience normal age-related attentional decline. Shneiderman (1991) points out the advantages and disadvantages of touch screens over other input devices. The advantages include:

- Touching a visual display of choices requires little thinking and is a form of direct manipulation that is easy to learn.
- Touch screens are the fastest pointing device.
- Touch screens have easier hand-eye coordination than mice or keyboards.
- No extra workspace is required as with other pointing devices.
- Touch screens are durable in public access and in high-volume usage.

The disadvantages include:

- Users' hands may obscure the screen.
- Screens need to be installed at a lower position and tilted to reduce arm fatigue.
- Some reduction in image brightness may occur.
- They cost more than alternative devices.

Although Shneiderman made these claims 20 years previously, they still apply to today's touch screen technology and the views are more recently echoed by Bhalla and Bhalla (2010). According to research these advantages outweigh the disadvantages for older users. For example, in a study comparing input devices for numeric entry tasks it was found that both younger and older participants preferred using the touch screen compared to physical keypad (Chung et al., 2010). Similarly, Umemuro (2004) showed that anxiety towards computers was reduced significantly for elderly users trained with a touch screen terminal, whereas there was no significant decline in anxiety for those using a keyboard based terminal. In comparison to a computer mouse, using a touch screen can significantly reduce the age effects for the time it takes to carry out pointing tasks (Iwase & Murata, 2002). Ease of use and reduced anxiety towards technology follow well designed systems and interfaces. In section 3.2 we will discuss some guidelines suggested for touch screen use.

3.2 Touch screen design guidelines

The usability website Sap Design Guild put together a comprehensive guide to designing touch screen applications (Waloszek, 2000; http://www.sapdesignguild.org/resources/

tsdesigngl/index.htm). In this document they state that there are five golden rules for designing touch screen interfaces. These include:

1. Speed - make sure the application runs fast,
2. Intuitiveness - avoid making the user think about what they need to do,
3. Choices - limit choices,
4. Guidance - guide the user through the application, and
5. Testing - use focus groups or observations to test the application before putting it in the "wild".

In relation to the interface design of touch sensitive displays, Waloszek (2000) also provided some guidance. The main areas that were covered were the screen layout, maintaining screen space, data entry, buttons and menus, and complex controls. We will outline some of the main points here along with other research conducted on these subjects.

3.2.1 Screen layout

Some of the factors that impact the experience for touch screen users are the size of the screen, the precision needed to select a target and the design of the screen elements (colour and grouping etc.). With large screens, designers can afford to offer large input and output features. Users should easily see menus and content on the one screen. On smaller screens however, such as mobile phones, there needs to be a compromise between these elements. To investigate just how much of a compromise is needed for older users Ziefle (2010) asked 40 participants between 55 and 73 years of age to take part in a series of navigation tasks. The focus of this work was between the visual density (font size; 8pt compared to 12pt) and preview size (menu; 1 option at a time compared to 5). Overall she found that the preview size had more impact on usability and navigation. Effectiveness and efficiency was lowest when only one menu option was shown on the screen, even with a large font size. The best performance was observed for the display with large font size and large preview.

With touch screen systems, users are not only looking at the screen, they are interacting with it. The layout of target buttons on the screen impacts this interaction. Larger features such as big buttons benefit users with vision problems but they also support accurate target selection. Conversely, decreasing the size of the buttons below finger width reduces performance (Lee & Zhai, 2009). Fitts (1954) hypothesized that the time needed to move to a target area is a function of the distance to the target and the size of the target. This speed/accuracy trade off is referred to as Fitts's Law and has been applied and modified extensively to assist interface designs such as target size and grouping, pop-up menus, and using the corners of the screen (Sears & Shneiderman, 1991). Providing large targets is particularly important for older users. Fezzani et al. (2010) looked at the effect of target size on performance for younger and older individuals. The study showed that reducing the target size resulted in difficulties with pointing accuracy, increased the time per task and increased the mental cost associated with performance. For the older users, the impact of motor difficulty on performance was considerable.

When designing the interface you need to consider what you want the user to see, for example, being consistent when grouping elements to aid navigation and eliminating extraneous features so the screen is not cluttered. Colour is also an important variable. For older or colour blind users it is necessary to have high colour contrast. Grouping menus by colour alone can also lead to difficulties for these users. Instead it would be preferable to use text, spacing or frames. Regarding the background colour, Waloszek (2000) does not recommend black or dark colours as they highlight finger prints and increase glare. Perhaps this suggestion is more applicable for public touch screen kiosks or applications for larger devices, as it contradicts the popularity of touch screen mobile phone applications such as on Apple's iPhone. It should be noted however that the iPhone offers a reversal of background/foreground colours as an accessibility option for those who want higher colour contrast (http://www.apple.com/accessibility/iphone/vision.html).

3.2.2 Data entry

Virtual alphabetic or numeric data entry is not ideal for touch screen users as the virtual keyboard or keypad obscures the screen and the inputting arm can quickly become fatigued. Data entry should be kept to a minimum, however if possible, other options could be offered, such as clicking on predefined values, providing buttons or sliders for increment and decrement values or lastly using a physical keypad or keyboard (Waloszek, 2000). Although there are issues with data entry on touch screens for extensive interaction, virtual data entry may be preferable to users for small data input activities. For example, as mentioned in section 3.1, both younger and older people show a preference for touch screen keypads in numeric entry tasks (Chung et al., 2010). As the virtual keypad is situated on the screen display the user does not have to divide their attention between the entry device and the screen content. This is of particular benefit to older users as there is an age-related decline in divided attention (see section 2.2).

Other types of input that have been considered as an input strategy are speech and eye-gaze. However, these forms of input are not commonly associated with touch screen systems. These methods would accommodate motor difficulties associated with ageing such as arthritis or tremble. Developments into these techniques are already underway. For example, Google have recently released an application which allows users to speak actions or search queries into their mobile phone (http://www.google.com/mobile/voice-actions/). However, there are several issues that impede the efficiency of speech input, such as the variation of users' intonation and the effect of external noise. Furthermore, Shneiderman (2000) pointed out in one of his articles "The Limits of Speech Recognition" that people have more difficulty simultaneously thinking and speaking, than they do thinking and walking, or thinking and typing on a keyboard. This is because the part of the brain responsible for problem solving and recall also supports speech.

Eye-gaze may have potential as an input strategy for older users. Murata (2006) examined pointing time and performance of young, middle aged and older individuals when using a mouse and also an eye-tracking system for input. Not surprisingly based on the previous research mentioned, compared to younger users the older adults found the mouse more difficult to use and were the slowest to perform the task. There were less age differences when using the eye-gaze input. Interestingly the younger participants liked the eye-gaze input least. This may be because they were more confident using the mouse. It was noted by the author that issues such as the necessity of calibration existed when using the eye-gaze input and that the technology would need to be further developed before it is considered as an acceptable input strategy.

3.2.3 Buttons and menus

Touch screen applications best afford point and select interaction and buttons are the most appropriate target for this action. The optimal button size for finger selection is believed to be at least 20mm square (Chung et al., 2010; Waloszek, 2000). Strauss (2009) proposed that wider buttons (20mm by 31.75mm) are more appealing to users based on the Golden Ratio Phi (Livio, 2002). Minimum recommended button size is 10mm (Lee & Zhai, 2009) and a slightly larger 11.43mm for older users (Jin et al., 2007). Regarding button spacing, Jin et al. recommended that a space between 3.17mm and 12.7mm is needed to lower performance error rates for older users.

The design elements of buttons, such as text, colour and icons, can either help or hinder users and should be considered carefully (Maguire, 1999). In a review of user guidelines for public touch screen kiosks Maguire reported on the necessity of providing text that users can see and read (such as sans-serif, 16pt), and terms and/or icons that users will recognise. It is also recommended that colour is used to group items together and provide activity feedback but that it should be used sparingly and with colour blind users in mind.

Physical hard buttons have advantages over soft touch screen buttons in that users have direct tactile feedback that the button has been pressed. With soft buttons on a screen, common forms of feedback are visual output such as changing the colour of the button after selection. Audio and haptic feedback, commonly associated with mobile phones, may benefit other forms of touch screen systems. For example, Lee and Zhai (2009) compared the speed and accuracy of users for soft buttons with no feedback, audio feedback, haptic (vibration) feedback, and both audio and haptic together. They found that having either audio or haptic significantly improved performance over having no feedback. There was no further increase for having both forms together however.

Menus provide a way for users to navigate a system. We mentioned previously that users perform better when they are displayed a large menu preview (Ziefle, 2010), or in other words, when they can see all their menu options. Maguire (1999) supports this finding stating that where possible all menu items should be displayed on the one page. Waloszek (2000) advises that different menu groups (e.g. main menu and secondary menu) should be differentiated by their shape and style.

3.2.4 Complex controls

Complex controls for touch screen applications include the use of lists and tables (Waloszek, 2000), which may be more suitable for high precision input such as a computer mouse. On standard PC's, scroll bars are typically used if the information extends beyond the screen size. However, scroll bars are not suited to touch screen interaction, and older users find them particularly difficult to use. Dickinson et al. (2010) carried out an 11 week study examining the barriers older people experience when learning to use computers. Some of the problems people found most difficult included double clicking the mouse and using scroll bars. As the researchers had ensured that all targets and icons were enlarged, there were no problems with precision or seeing objects on the screen. Features such as double tapping and scroll bars should be avoided for touch screen use to accommodate older users.

Other options to view an extended screen could be incorporating next/previous or up/down buttons, or using gestures such as swiping the screen horizontally or vertically, depending on the navigation structure. Swiping gestures are common forms of input for smaller touch screen systems such as mobile phone or tablets where the user can swipe their index finger from one side of the screen to another to change the page. Larger surfaces would require more physical effort for this task to be achieved. Familiar gestures such as using a finger or stylus in a crossing off motion (X) to delete an item or ticking (✓) to save an item are particularly compatible for older users learning this input strategy (Stöbel, 2009).

3.3 Designing for disability

The purpose of this chapter is to discuss the accessibility of touch screen technology for older users and to review the recommendations for designing touch screen interfaces. This is with the assumption that both able-bodied users and users with impairments can benefit from touch screen devices like mobile phones, tablets and public kiosks. Newell (2008) argues however that the concept of "design for all" is unrealistic as there are distinctions between mainstream technology and assistive technology (including rehabilitative devices). Newell outlines some of the factors that should be considered when designing for users with a disability:

- Much greater variety of user characteristics and functionality.
- The difficulty in finding and recruiting "representative users".
- Possible conflict of interest between accessibility for people with difference types of disability.
- Conflicts between accessibility, and ease of use for less disabled people.
- Situations were "design for all" is certainly not appropriate (e.g. blind drivers of motor cars)
- The need to specify exactly the characteristics and functionality of the user group.
- Provision of accessibility via the provision of additional components.

Although Newell makes the point that users with disabilities have specific needs when it comes to technology, he points out that the aesthetics of a device are just as important as its functionality (Newell, 2003; Newell, 2008). He reasons that assistive technology could look more like the aesthetically pleasing mainstream technology if developers were motivated enough. There has been some improvement over the last few years suggesting that developers are listening to their users. For example, current hearing aids are much more discrete compared to past designs.

Often, older and disabled users come under the same category as both groups tend to have one or more perceptual, cognitive or physical disabilities. In the previous section we have discussed the design needs of older users in relation to target size and spacing etc. but it is interesting to consider whether users with disabilities have similar needs. To answer this question, Irwin and Sesto (2009) examined the performance of people with and without motor control disabilities on a touch screen device. Of the 30 participants in the study, there were 11 with Cerebral Palsy, 11 with Multiple Sclerosis and 8 age matched controls. Performance levels were based on timing and accuracy using buttons ranging in size from 10mm square to 30mm square, with spacing ranging between 1mm and 3mm. Similar to older users, the optimum button size for the participants in this study was between 20 and

25mm. There was a 23% reduction in performance time for buttons at size 25mm and error rates began to level off for buttons with size 20mm. If the screen size is large enough, designers could easily accommodate for disabled users by implementing these features in application interfaces or at the very least implementing the option to adjust settings to increase accessibility.

Mobile touch screen interfaces are more challenging due to the limited screen space. Besides the applications installed in touch screen mobile phones, there are many usability benefits for people with motor impairments. For example, the soft buttons support users with low muscle strength and the phone itself can be easily carried and used in one hand. Touch screens devices afford many types of user input. Users can tap the screen to select a target, swipe across the screen to view up or down a page, and they can resize the screen by pinching or releasing their fingers on the screen surface. Guerreiro et al. (2010) evaluated touch techniques with a group of tetraplegic participants. They looked at the techniques of tapping, crossing (swiping), exiting (swiping the edge or corners to leave the screen) and directional gesturing (swiping in a particular direction). The participants were asked to complete tasks using these input gestures. The size and position of targets were also under scrutiny. There were several interesting observations from this study. Firstly there was little difference in performance between the medium (12mm) and large (17mm) target size. There was a significantly higher error rate for the smallest targets (7mm), supporting Lee & Zhai's (2009) finding that using a target size below 10mm decreases performance. The position of a target (corner, edge or middle of the screen) did not affect performance however the authors suggest that using corners and edges allow participants to tap targets more precisely. The exiting gesture proved to be the most difficult whereas tapping and crossing gestures produced higher performance levels. These findings should give designers some insight into appropriate target design and gesture input for motor impaired users.

Touch screen devices do not offer the tactile feedback that people with visual impairments rely on with physical objects, like a keyboard or telephone keys. Screen readers provide blind and visually impaired computer users access to information on their screen. VoiceOver is an application available on the iPhone which tells the user what item they are currently touching (http://www.apple.com/accessibility/iphone/vision.html). The speed of the voice over can be adjusted depending on the users preference and other phone related sounds are automatically lowered when the voice over is activated. There is no doubt that this application is a useful one. Bonner et al. (2010) evaluated the usability of iPhone's VoiceOver text entry system compared with an eyes-free technique that they developed called No-Look Notes. The aim of the No-Look Notes technique was to provide a text entry system for blind users that was robust, had a familiar layout and allowed users to painlessly explore the system's layout. Keyboard characters are arranged in an 8 segment pie menu from which the user can select one segment (e.g. ABC) (see figure 3). This brings them to the secondary screen where they can select their letter of choice (e.g. A). Different gestures, such as resting a finger on a segment and tapping a second finger, are used to allow the user to both explore and select with their fingers. The authors noted that they did not integrate haptic (vibration) feedback into the system as this may not be available on all mobile phones. Visual and auditory feedback were included however. An iPhone was used to evaluate both text entry techniques, VoiceOver and No-Look Notes. A group of 10 visually impaired participants were asked to use both systems for one hour (after 15 minutes practice), inputting words from a predefined list. Performance was measured based on entry speed and accuracy. A questionnaire was also given at the end of each session to obtain subjective feedback. Nearly all of the participants (90%) performed significantly faster using the No-Look Notes system and error rates were lower. Although participants were enthusiastic about both techniques, they rated the No-Look Notes as easier to use and learn.

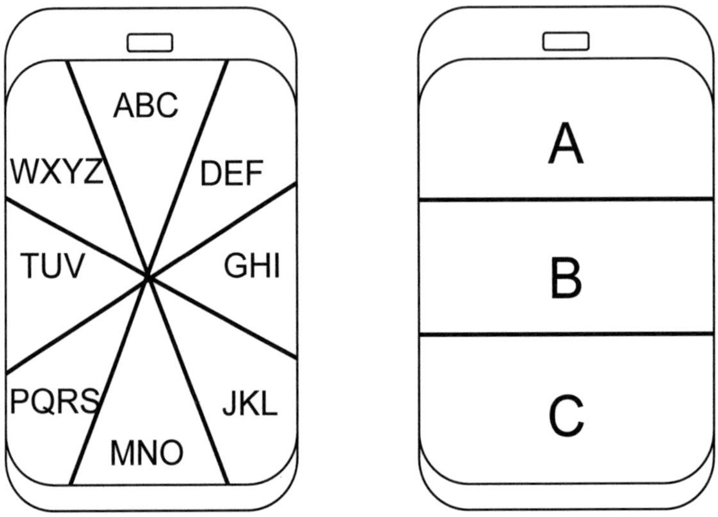

Fig. 3. Structure of touch screen keyboard for No-Look-Notes (Bonner et al., (2010); Left image shows the main screen from which the user selects a letter grouping. This allows them to then choose a specific letter, as shown on the right.

Raman and Chen (2010) at Google also created a method for eyes-free gesture input on a touch screen mobile phone. This method was designed to support people to accurately enter characters without having to see the screen. Instead of the users trying to find the correct character, wherever they touch on the screen becomes 5. The user can then swipe in different directions to enter further numbers, up for 2, down for 8 etc. The input is also supported by visual, auditory and haptic feedback. Having multiple forms of feedback like this is useful for users of all abilities to compensate for noise or distractions in the environment.

In this section we have outlined some of the design recommendations for touch screens put forward by researchers and designers in the field. Adhering to these guidelines is particularly important for older and disabled users who may have perceptual, cognitive, and physical difficulties.

4. Touch screen applications

We have touched on some of the research which highlighted the benefit of touch screens for older users compared with other input devices. It is because of these advantages that touch screen technology is so frequently used for general public use. The following are only some of the applications of touch screens.

4.1 Public environments

Public information displays must cater for the largest possible target group, including the young, old, disabled and non-native. Public touch screen kiosks include airport self-check in, supermarket self-checkout, tourism information points, public transport ticket dispensers and many more. In many retail establishments staff also use touch screen devices. To accommodate the broad range of users these public touch screens tend to have large screen displays at a low height or with a tilted screen. The interface usually displays large target options with a combination of image and text, along with short concise instructions (see figure 4 for example). This allows for information to become easily and quickly accessible for users only requiring finger-touch interaction.

Fig. 4. Airport check-in kiosk; The interface makes use of large, short instructions and combines text labels with images for target options, supporting non-English speaking users.

In order for touch screen kiosks to be attractive to the public, it should be quicker and easier to use this system compared to a person serviced desk. This is not always the case however. In supermarkets for example, self-checkout systems are often a cause of frustration for customers (Smith, 2010). Although the self-checkout can be quicker for customers with a small number of items, often the attention of staff is needed to verify the age of a customer buying alcohol, to remove security tags or if there are any errors messages. Until a more efficient system is put into place for shoppers it is likely that the preference of customers will remain with manned tills. Public kiosks also have to take into account the context of where it will be used. The location of public kiosks needs to be conspicuous so that people will notice and use them. Applications need to be very easy to use to accommodate for the busy and noisy environments where they are situated, such as supermarkets and airports. The scenario of desperately trying to figure out how to use a self-service system while a queue forms behind you is familiar for many people, young and old. The pressure is amplified if the people in the queue are in a hurry, for example to catch the next train or flight. It is also particularly important that users feel competent interacting with systems such as ATMs that might require the input and output of private information. Systems such as these can be placed in a secluded area such as a booth, so that users feel more secure about entering sensitive information in a public place. This can also help to reduce the exposure to noise and other distractions.

4.2 Home environments

The integration of touch screen systems into the home environment has been somewhat slower compared to public and mobile phone touch screen use. The popularity of tablet computers, coupled with the advances made in sensor network monitoring, has encouraged the development of a series of applications based on home living. For example, the home security company uContol recently released a touch screen system which can be placed in the user's home allowing them to monitor and control the security status of the house (http://www.youcontrol.com/touchscreen.html). Another example of home monitoring is in relation to energy consumption displays. Such prototypes include home energy systems developed by Intel (cited in Eisenberg, 2010) and CLARITY (see figure 6; Doherty et al., 2010).

Researchers and manufacturers have taken note of this touch screen "revolution" and it might not be long before we see touch sensitive displays embedded in many of our home appliances. For example, Gorenje introduced the iChef, an oven with a large touch screen colour display, where the user can choose different functions such as the temperature, timer control etc. (http://www.gorenjegroup.com/en/livingkitchen2011-pressroom/ichef-revolutionary-oven-touch-control). Similarly, Luo, Jin, and Li (2009) proposed the concept of a "smart fridge" which consisted of a touch screen display integrated into the door of a fridge. The purpose was to offer nutritional and recipe advice based on the food contained in the fridge.

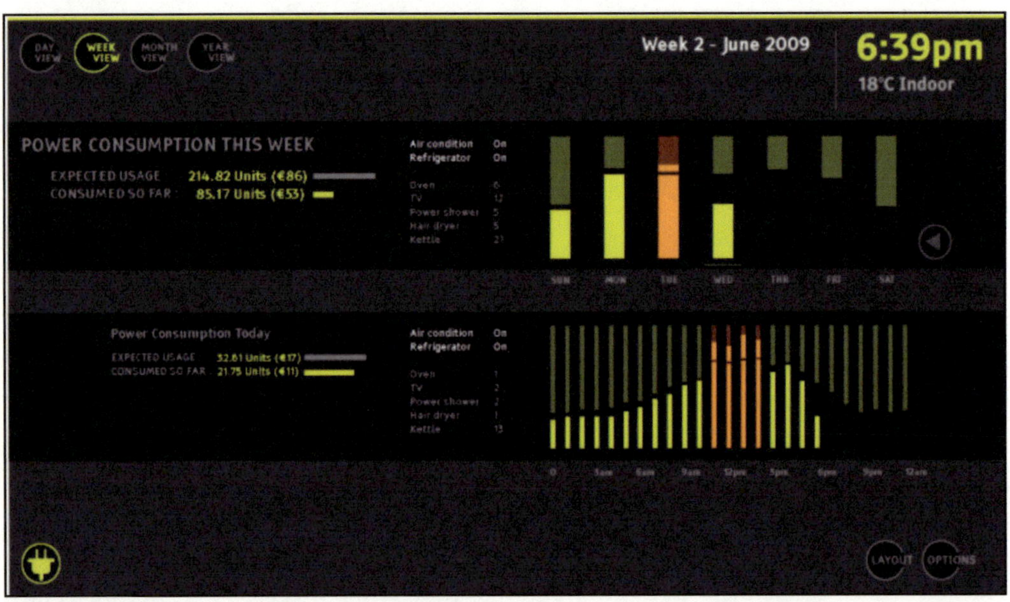

Fig. 5. Example of dark background with bright information; In-home display electricity monitor, Doherty et al. (2010).

It is worth noting that all of the touch screen interfaces mentioned in this section have one similar design characteristic; the colour scheme. In all of these interfaces a dark grey and black background is used, with important information highlighted in a bright colour and less pertinent information displayed in a light grey colour (see figure 5 for example). A reason for this display option is given by Doherty et al. (2010). They explain that having a dark background and bright information allows the display to be seen in a bright or darkly lit room. For example, a bedside digital clock can be easily read in a dark room, however if the background display was white it would momentarily blind the sleepy individual.

There are obvious benefits to applications such as the ones just mentioned, including security, and awareness of energy consumption etc. However it is not only the usefulness and usability of the device that is significant. Factors such as the cost of the device, the Internet connection and the subscription to the company monitoring the home have to be considered, particularly in the context of older users who may not have a regular income. These issues may not be a problem in the future if technology costs are reduced and applications like the ones mentioned above come as a standard feature of household security and electricity services.

4.3 Healthcare environments

There has been extensive research carried out looking at the use of touch screens in medical environments. This may be due to the busy nature of hospitals and medical centres where staff cannot always provide patients with the amount of time they would like. Medical centres such as the Summer Town Health Centre in Oxford, UK, have installed a touch screen arrival system for patients to register their information before an appointment. Other applications include touch screen way-finding kiosks in hospitals (Wright et al., 2010), medication management applications (Siek et al., 2011), and self-health status assessment (Yost et al., 2010).

Many studies using touch screen kiosks do not assess participants' subjective opinions about the touch screen technology, concentrating more on the applications installed. Wright et al. (2010) for example noted that during their ethnographic study of a hospital way-finding touch screen kiosk, it was impractical to interview visitors as they were in a hurry to get to their destination. Other studies were able to obtain feedback however. For instance, Nicolas et al. (2003) found that a third of people using a touch screen health information kiosk found it difficult to use compared to only 23% of Internet users. They believed that the kiosk users were less computer literate and therefore less familiar with the computer terms such as "prev screen" used in the application. This highlights the importance of following design guidelines to support users of all levels. In contrast to this finding, Allenby et al. (2002) looked at the use of touch screen technology for obtaining self-report information from 450 cancer patients. The touch screen systems were placed in cubicles to support patient privacy. Accessibility was accommodated for patients in wheelchairs or confined to a bed by providing mobile touch screen systems. An assistance button was also included on the screen so that users could call a nurse if they needed help. After the participants completed the touch screen self reports, they were asked to fill in a questionnaire regarding touch screen acceptability. Overall, 99% of the participants found the touch screen easy to use and over half of these had never used a computer before.

Way-finding in hospital buildings can be challenging for visitors. This may be because of the large size of the buildings and the division into clinical departments and wards. The state of mind of the person visiting the hospital should also be taken into account, as it is likely they will be upset or in a hurry to find their sick friend or relative. To compensate for this problem, Wright et al. (2010) developed a touch screen kiosk for visitors to retrieve way-finding information. Particular attention was paid to the accessibility of the kiosk. This was in relation to the physical structure, the interface and the information output. For example, the kiosk was put at wheelchair height at a slant so that standing users would be able to use it easily. The interface consisted of a limited number of large targets, an audio output option and an animated map with a photograph of the destination. Users were also provided with the option to find wheelchair friendly routes avoiding staircases. Wright et al. carried out a field study over 10 weeks to evaluate the usefulness of this system. During this time they followed 22 visitors who used the kiosk and they found that 19 of the 22 successfully reached their destination without further help. However, hospital staff reported that after using the system visitors often asked for further way-finding information at the reception desk. Feedback concerning the usability of the system could not be obtained as the users were in a hurry to find their destination; therefore, it was unclear whether the users had difficulty using the touch screen kiosk itself or whether they were looking for reassurance about the directions given.

Health care systems do not have to be limited to hospital or health-based information. Bedside patients can also benefit from touch screen applications to entertain them and alleviate boredom. An example of this is MEDIVista, an innovative computing platform developed by an Irish bedside computing and media solutions company, Lincor Solutions (Murphy & O'Donnell, 2010). MEDIVista is an information management and entertainment system placed at a patient's hospital bedside. The terminal comprises of a touch screen and telephone mounted on a swivel arm for easy patient access. The patient information system consists of electronic hospital information, patient records, meal ordering, and employee tracking. The entertainment system comprises of a television, radio, Internet, audio books, online shopping, games, and films on demand. Nearly 20,000 of these systems are currently installed in hospitals situated in Europe, North America and Asia Pacific. The widespread distribution of MEDIVista suggests that it is a well received application by both hospital staff and patients.

Touch screen applications in health care settings can be a valuable time saver for staff or a source of entertainment for patients. The key factors that need to be taken into account for these applications are (1) accessibility - in a hospital setting it is likely that users will not have full physical abilities or may be wheelchair or bed bound, (2) knowledge of computer terms - users, particularly older people, may have never used a computer before and would not be familiar with computer-related terms, and (3) knowledge of health terms - users may not be familiar with medical terminology, therefore they should be explained clearly. From the studies outlined above it is clear that although touch screen systems can support hospital patients and visitors up to a point, most often people in healthcare environments need comfort and advice along with medical information that only humans can provide.

4.4 Communication and social devices

Communication devices such as mobile phones are probably the most popular touch screen for social use. However, research has shown that older adults are less likely to use mobile phones than younger adults (Cullen et al., 2008). Kurniawan (2008) examined older adults' mobile phone use and found that it was mostly limited to basic functions such as making calls and sending text messages. In this study, older participants claimed that the design of mobile phones, such as the size of screen features, inhibited their use of the device. The mobile phones discussed in these studies were not touch screen enabled, however the issues raised are common problems and can be applied generally.

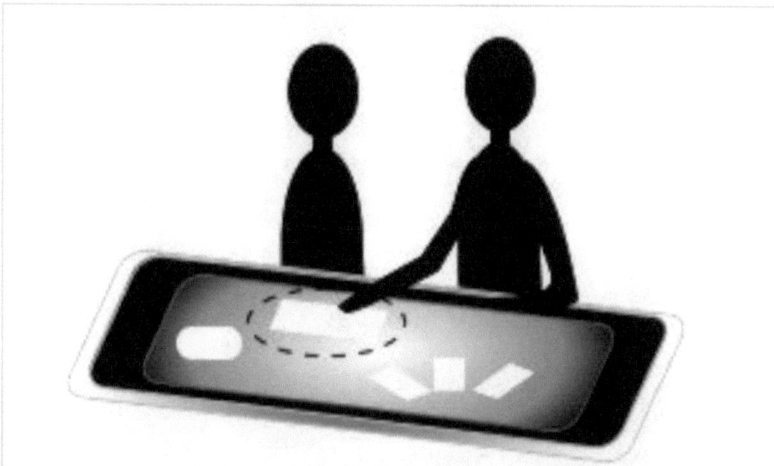

Fig. 6. Multi-touch screen tabletops allow multiple users to interact with the data on display.

Novel applications have been developed for mobile phones to encourage social interaction of multiple users with the one device. A storybook application, StoryKit, was designed for the Apple iPhone (Quinn et al., 2009). With this application story books are created by inputting text, illustrations, audio recordings and photographs into the device. The purpose of this was to promote adult guided learning, encouraging grandparents and grandchildren to work together to create stories on the mobile phone. Although the creative and social aspect of the application was well liked by both older and younger users, some older users reported difficulty using intricate iPhone features such as the touch screen keyboard. Although using a touch screen phone as a platform for socially interactive applications supports the mobile aspect of using features in the environment to build a story, larger screen sizes may be more suitable for multi-person usage.

An example of a social interaction system on the opposite end of the size scale is Sharetouch (Tsai et al., 2011), a multi-touch tabletop (see figure 6). The purpose of this system is to act as a coffee table and share-display system in one, so that older adults in senior centres can communicate with others, share data, or play games. Other systems have been designed to support reminiscence between older adults and their family, friends or caregivers. The CIRCA reminiscence system was designed to support communication and reminiscence between older adults with dementia and their carers (Astel et al., 2010). The system is a touch screen device that allows users to explore a set of text, photographs, sound recordings and film recordings. These media items act as a trigger for discussion and allow the older user control over the discussion topic. Astel et al. found that interacting with the CIRCA system increased the level of eye contact and intimacy between older participant and their carer.

Touch screen technology could also support older adults' lifelogging activities. Lifelogging is automatically recording activities and events in one's life and is now becoming more and more possible with sensors such as accelerometers (recording movement), Bluetooth (identifying the people around you) and cameras (capturing a visual record) being

Fig. 7. Touch screen to support older adults browsing through lifelog images (Caprani et al., 2010).

integrated into everyday devices like mobile phones. Recording all this information can provide people with a rich collection of information about what they did, where they were and who they were with. Even after only a year, a person could collect millions of files and presenting this vast amount of data in a manageable way is challenging. However researchers have come up with methods to do this, such as segmenting the information into separate events or activities (Doherty, 2009). Using this segmentation method, Caprani et al. (2010) designed a touch screen browser for older users to upload and view lifelog images (see figure 7). Older participants were involved throughout the design process, advising features such as text and image size, colour contrast and labelling terms. Each image on the main screen represents an event. An event is a group of images related to one activity like eating lunch or doing the crossword. A field study was conducted over a period of two weeks to evaluate the usability of the browser with a small sample of older adults with no computer experience. Although there were initial difficulties learning to turn on and off the computer and double tapping to open the browser, overall the participants felt comfortable using the touch screen. They also reported feeling more confident towards computers and enjoyed interacting with it. It has been shown that lifelog collections can support recall for people with memory impairments (Berry et al., 2009). Therefore a lifetime of data representing our activities could support our memory as we grow older. Until then, other motivations for capturing visual lifelogs could be to support reminiscence by sharing and exchanging these photographs with family (Caprani et al., 2011) or using the data to create narratives of events that we experienced in our life (Byrne et al., 2011).

The purpose of social and communication devices is to encourage individuals to interact with each other through a medium. Touch screen devices offer older users with limited computer experience access to this. In the examples outlined above, the ease of use of the systems allowed the older users to take control of their own activities and conversations with others. Communication technology can help to alleviate the feeling of isolation and loneliness that many older adults experience by providing them access to friends and family. This may be through phone or video calls, email or data sharing.

5. Conclusion

Shneiderman's prediction in 1991 that touch screen technology would become ubiquitous has proven to be an acutely accurate one. We have discussed in this chapter the progress that has been made since touch screens were first conceptualised by E.A. Johnson in 1965. Now touch screen technology is used in public, healthcare, and home environments, and the commercial market has greatly profited from the popularity of touch screen mobile phones and tablet computers. But it is the intuitive nature of direct finger input that makes touch screen technology so appealing to users of all ages and abilities.

An increase in longevity together with declines in birth-rate has resulted in an ageing population and developers have come to realise that older adults will soon be a major user market. Technology is becoming easier to use and many systems incorporate accessibility features to support users with impairments. Research has shown that older adults perform better and prefer using touch screens compared to other input devices. There is also less of an age-related difference in performance when using touch technology. Having said this there is still some way to go before mainstream technology can fully accommodate for the needs of older and disabled users. However, as we can see from the research outlined in this chapter, there has been extensive work carried out investigating optimum design features for touch screen systems to support older and disabled users. These have included button size and spacing, menu structures, and data entry among others. Although other forms of input such as eye-gaze or indeed the computer mouse may become more usable for older adults in the future, touch screens provide an accessible alternative. By following the guidelines and recommendations put forward by researchers in this field it should now be possible to design applications that are functional, accessible and aesthetically beautiful.

6. Acknowledgements

The funding for this research has been provided by the Irish Research Council for Science, Engineering and Technology. This work is also supported by Science Foundation Ireland under grant 07/CE/I1147.

7. References

Allenby, A., Mathews, J., Beresford, J., & McLachlan, S.A. (2002). The Application of Computer Touch-Screen Technology for Psychosocial Distress in an Ambulatory Oncology Setting. *European Journal of Cancer Care,* Vol.11, No.4, (December 2002), pp. 245-253

Astell, A.J., Ellis, M.P., Bernardi, L., Alm, N., Dye, R., Gowans, G., & Campbell, J. (2010). Using a Touch Screen Computer to Support Relationships between People with

Dementia and Caregivers. *Interacting with Computers*, Vol.22, (July, 2010), pp. 267-275, ISSN 0953-5438

Berry, E., Hampshire, A, Rowe, J., Hodges, S., Kapur, N., Watson, P., Browne, G., Smyth, G., Wood, K. & Owen, A. (2009). The Neural Basis of Effective Memory Therapy in a Patient with Limbic Encephalitis. *Journal of Neurology, Neurosurgery, and Psychiatry with Practical Neurology*, Vol.80, No.11, (November 2009), pp. 1202-1205

Bhalla, M.R., & Bhalla, A.V. (2010). Comparative Study of Various Touchscreen Technologies. *International Journal of Computer Applications,* Vol.6, No.8, (September 2010), pp. 12-18, ISSN 09758887

Bonner, M.N., Brudvik, J.T., Abowd, G.D., & Edwards, W.K. (2010). No-Look Notes: Accessible Eyes-Free Multi-Touch Text Entry, *Proceedings of Pervasive Computing 2010*, pp. 409-426, ISBN 978-3-642-12653-6, Helsinki, Finland, May 17-20, 2010

Bosman, E.A. & Charness, N. (1996). Age Differences in Skilled Performance and Skill Acquisition, In: *Perspectives on Cognitive Change in Adulthood and Aging,* T. Hess & F. Blanchard-Fields, 428-453. McGraw-Hill, ISBN 0070284504, New York

Bouwhuis, D.G. (2003). Design For Person-Environment Interaction in Older Age: A Gerontechnological Perspective. *Gerontechnology,* Vol.2, No.3, (March 2003), pp. 232-246

Brown, D., Steinbacher, C., Turpin, T., Butler, R., & Bales, C. (2011). History of Elo, In: *Elo TouchSystems,* 26.05.2011. Available from http://www.elotouch.com/AboutElo/History/default.asp

Byrne, D., Kelliher, A., & Jones, G. (2011). Life Editing: Third-Party Perspectives on Lifelog Content, *Proceedings of Human Factors in Computing Systems (CHI '11),* pp. 1501-1510, ISBN 978-1-4503-0228-9, Canada, May 7-12, 2011

Caprani, N., Doherty, A.R., Lee, H., Smeaton, A.F., O'Connor, N.E., & Gurrin, C. (2010). Designing a Touch-Screen SenseCam Browser to Support an Aging Population, *Proceedings Human Factors in Computing Systems (CHI EA '10),* pp. 4291-4296, ISBN 978-1-60558-930-5, Atlanta, USA, April 10-15, 2010

Caprani, N., O'Connor, N.E., & Gurrin, C. (2011). Motivating Lifelogging Practices through Shared Family Reminiscence, *Proceedings of Human Factors in Computing Systems (CHI EA'11),* pp. 10-15, ISBN 978-1-4503-0268-5, Vancouver, Canada, May 7-12, 2011

Charness, N. & Boot, W.R. (2009). Aging and Information Technology Use: Potential and Barriers. *Current Directions in Psychological Science,* Vol.18, No. 5, (October, 2009), pp. 253-258, ISSN 09637214

Charness, N., & Jastrzembski, T.S. (2009). Gerontechnology, In: *Future interaction design II,* P. Saariluoma & H. Isomaki (Eds.), 1-29, Springer-Verlag, ISBN 978-1-84800-300-2, London

Chung, M.K., Kim, D., Na, S., & Lee, D. (2010). Usability Evaluation of Numeric Entry Tasks on Keypad Type and Age. *International Journal of Industrial Ergonomics,* Vol.40, No.1, (January 2010), pp. 97-105, ISSN 0169-8141

Coleman, G.W., Gibson, L., Hanson, V.L., Bobrowicz, A., & McKay, A. (2010). Engaging the Disengaged: How Do We Design Technology for Digitally Excluded Older Adults? *Proceedings of Designing Interactive Systems (DIS '10),* pp. 175-178, ISBN 978-1-4503-0103-9, Denmark, August 16-20, 2010

Cullen, K., Dolphin, C., Delaney, S., & Fitzpatrick, M. (2008). Survey of Older People and ICTs in Ireland, In: *Age Action*, 27.04.2010, Available from: www.ageaction.ie/userfiles/.../survey-of-older-people-andicts-in-ireland.pdf

Czaja, S.J. & Lee, C.C. (2007). Information Technology and Older Adults, In: *Human-Computer Interaction Handbook: Fundamentals, Evolving Technologies and Emerging Applications (2nd Edition)*, A. Sears & J. A. Jacko, 777-792, Erlbaum, ISBN 0805858709, New Jersey, USA

Czaja, S.J., Charness, N., Fisk, A.D., Hertzog, C., Nair, S.N., Rogers, W.A., & Sharit, J. (2006). Factors Predicting the Use of Technology: Findings from the Center for Research and Education on Aging and Technology Enhancement (CREATE). *Psychology and Aging*, Vol. 21, No.2, (June 2006), pp. 333-352

Dickinson, A., Eisma, R., & Gregor, P. (2010). The Barriers that Older Novices Encounter to Computer Use. Universal Access in the Information Society, Vol.10, No.3, (September 2010), pp. 261-266

Doherty, A.R. (2009). *Providing Effective Memory Retrieval Cues through Automatic Structuring and Augmentation of a Lifelog of Images.* Unpublished doctoral dissertation, Dublin City University, Dublin, Ireland

Doherty, A.D., Qiu, Z., Foley, C., Lee, H., Gurrin, C., & Smeaton, A.F. (2010). Green Multimedia: Informing People Of Their Carbon Footprint Through Two Simple Sensors. *Proceedings of Multimedia (MM '10)*, pp. 441-450, ISBN 978-1-60558-933-6, Firenze, Italy, October 25-29, 2010

Eisenberg, A. (February 2010). Energy Scoreboards, Designed for the Home, In: *The New York Times*, 17.07.2011. Available from http://www.nytimes.com/2010/02/28/business/28novel.html

Fezzani, K., Albinet, C., Thon, B., & Marquie, J. (2010). The Effect of Motor Difficulty on the Acquisition of a Computer Task: A Comparison between Young and Older Adults. *Behaviour & Information Technology*, Vol. 29, No.2, (March 2010), pp. 115-124

Fisk, A., Rogers, W.A., Charness, N., Czaja, S.J., & Sharit, J. (2009). Designing for Older Adults: Principles and Creative Human Factors Approaches (2nd Edition), CRC Press, ISBN 978-1420080551, New York

Fitts, P.M. (1954). The information capacity of the human motor system in controlling the amplitude of movement. *Journal of Experimental Psychology*, Vol. 47, No. 6, (June 1954), pp. 381–391

Guerreiro, T., Nicolau, H., Jorge, J., & Gonçalves, D. (2010). Towards Accessible Touch Interfaces, *Proceedings of ACM SIGACCESS conference on Computers and Accessibility (ASSETS '10)*, pp. 19-26, ISBN 978-1-60558-881-0, Orlando, Florida, October 25-27, 2010

Hawthorn, D. (2000). Possible Implications of Aging for Interface Designers. *Interacting with Computers*, Vol.12, No.5, (April 2000), pp. 507-528

Irwin, C.B., & Sesto, M.E. (2009). Timing and Accuracy of Individuals With and Without Motor Control Disabilities Completing a Touch Screen Task, *Proceedings of Universal Access in Human-Computer Interaction (UAHCI '09)*, pp. 535 – 536, ISBN 978-3-642-02709-3, Calafornia, USA, July 19-24, 2009

Iwase, H., & Murata, A. (2002). Empirical Study on the Improvement of Usability for Touch-Panel for Elderly: Comparison of Usability between Touch-Panel and Mouse,

Proceedings of Systems, Man and Cybernetics, pp. 252 - 257, ISBN 0-7803-7437-1, Hammamet, Tunisia, October 6-9, 2002

Jay, G.M., & Willis, S.L. (1992). Influence of Direct Computer Experience on Older Adults' Attitudes towards Computers. *Journal of Gerontology: Psychological Sciences,* Vol.47, No.4, (July 1992), pp. 250-257

Jin, Z.X., Plocher, T., &. Kiff, L. (2007). Touch Screen User Interfaces for Older Adults: Button Size And Spacing, *Proceedings of Universal Access in Human Computer Interaction: Coping with Diversity (UAHCI '07),* pp. 933-941, ISBN 978-3-540-73278-5, Beijing, China, July 22-27, 2007

Johnson, E.A. (1965). Touch Display: A Novel Input/Output Device for Computers. *Electronics Letters,* Vol.1, No.8, (October 1965), pp. 219–220

Kester, J.D., Benjamin, A.S., Castel, A.D., & Craik, F.I.M. (2002). Memory in Elderly People, In: *The Handbook of Memory Disorders,* A.D. Baddely, M.D. Kopelman, B.A. Wilson (Eds.), 543-568, Wiley, ISBN 047149819X, London

Kurniawan, S. (2008). Older people and Mobile Phones: A Multi-Method Investigation. *International Journal of Human-Computer Studies,* Vol.66, No. 12, (December 2008), pp. 889-901, ISSN 1071-5819

Lee, S., & Zhai, S. (2009). The Performance of Touch Screen Soft Buttons, *Proceedings of Human Factors in Computing Systems (CHI '09),* pp.309-318, ISBN 978-1-60558-246-7, Boston, USA, April 4-9, 2009

Livio, M. (2002). The Golden Ratio: The Story of Phi, the World's Most Astonishing Number. Broadway Books, ISBN 0-7679-0815-5, New York, USA

Luo, S., Jin, J.S. & Li, J. (2009). A Smart Fridge With an Ability to Enhance Health and Enable Better Nutrition. International Journal of Multimedia and Ubiquitous Engineering, Vol.4, No.2 (April 2009), pp. 69-80

Maguire, M.C. (1999). A Review of User-Interface Design Guidelines for Public Information Kiosk Systems. *International Journal of Human-Computer Studies,* Vol.50, No.3, (March 1999), pp.263-286

Mamolo M., & Scherbov S. (2009). Population Projections for Forty-Four European Countries: The Ongoing Population Ageing. In: *European Demographic Research Papers,* 02.08.2011, Available from http://www.oeaw.ac.at/vid/download/edrp_2_09.pdf

McLaughlin, A.C., Rogers, W.A., & Fisk, A.D. (2009). Using Direct and Indirect Devices: Attention Demands and Are-Related Differences. *ACM Transactions on Computer-Human Interaction,* Vol.16, No.1, (April 2009), pp.1-15

Mitzner, T.L., Boron, J.B., Fausset, C.B., Adams, A.E., Charness, N., Czaja, S. J., Dijkstra, K., et al. (2010). Older Adults Talk Technology: Technology Usage and Attitudes. *Computers in Human Behavior,* Vol.26, No.6, (November 2010), pp.1710-1721, ISSN 07475632

Morris, A., Goodman, J., & Brading, H. (2007). Internet Use and Non-Use: Views of Older Users. *Universal Access in the Information Society,* Vol.6, No.1, (May 2007), pp. 43-57, ISSN 16155289

Murata, A. (2006). Eye-Gaze Input versus Mouse: Cursor Control as a Function of Age. *International Journal of Human-Computer Interaction,* Vol.21, No.1, (January 2006), pp. 1-14, ISSN 10447318

Murphy, E., & O'Donnell, P. (2010). U.S. Patent No. 0017437. Washington, DC: U.S. Parent and Trademark Office

Newell, A. (2003). Inclusive Design or Assistive Technology, In: *Inclusive Design – Design for the Whole Population*, J. Clarkson, S. Keates, & C. Lebbon (Eds.), 172-181, Springer, ISBN 978-1852337001, London

Newell, A. (2008). User Sensitive Design for Older and Disabled People, In: *Technology for Aging, Disability and Independence: Computer and Engineering Design and Applications*, A. Helal, M. Mokhtari, B. Abdulrazak (Eds.), 787-802, John Wiley & Sons, Inc, ISBN 9780471711551, New Jersey

Nicolas, D., Huntington, P., & Williams, P. (2003). Delivering Consumer Health Information Digitally: A Comparison Between the Web and Touchscreen Kiosk. *Journal of Medical Systems*, Vol.27, No.1, (February 2003), pp. 13-34

Quinn, A. J., Bederson, B. B., Bonsignore, E. M., Druin, A. (2009). Storykit: Designing a Mobile Application for Story Creation by Children and Older Adults, In: *Technical Report HCIL-2009-22*, 01.08.2011, Available from http://hcil.cs.umd.edu/trs/2009-22/2009-22.pdf

Raman, T.V., & Chen, C.L. (2010). U.S. Patent No. 0073329. Minneapolis, MN: U.S. Patent and Trademark Office

Rogers, W.A., Fisk, A.D., McLaughlin, A.C., & Pak, R. (2005). Touch a Screen or Turn a Knob: Choosing the Best Device for the Job. *Human Factors*, Vol.47, No.2, (June 2005), pp. 271-88

Sears, A., & Shneiderman, B. (1991). High Precision Touchscreens: Design Strategies and Comparisons with a Mouse. *International Journal of ManMachine Studies*, Vol.34, No.4, pp. 593-613, ISSN 00207373

Selwyn, N. (2004). The Information Aged: A Qualitative Study of Older Adults' Use of Information and Communications Technology. *Journal of Aging Studies*, Vol.18, No.4, (November 2004), pp. 369-384, ISSN 0890-4065

Shneiderman, B. (1991). Touch Screens Now Offer Compelling Uses. *Software, IEEE*, Vol.8, No.2, (March 1991), pp. 93-94, ISSN 0740-7459

Shneiderman, B. (2006). The Limits of Speech Recognition. *Communications of ACM*, Vol.43, No.9, (September 2000), pp. 63-65

Siek, K.A., Khan, D.U., Ross, S.E., Haverhals, L.M., Meyers, J., & Cali, S.R. (2011). Designing a Personal Health Application for Older Adults to Manage Medications: A Comprehensive Study. *Journal of Medical Systems*, DOI 10.1007/s10916-011-9719-9

Smith, L. (August 2010). Self-Service, or Merely Self-Serving? The Revolution at the Tills, In: *The Independent*, 01.08.2011, Available from http://www.independent.co.uk/news/uk/home-news/selfservice-or-merely-selfserving-the-revolution-at-the-tills-2059363.html

Stöbel, C. (2009). Familiarity as a Factor in Designing Finger Gestures for Elderly Users, *Proceedings of Human-Computer Interaction with Mobile Devices and Services (MobileHCI '09)*, ISBN 978-1-60558-281-8, Bonn, Germany, September 15-18, 2009

Strauss, M. (August 2009). Touch Screen Button Dimensions and Spacing: Hell if Phi Know, In: *Home Technology E-Magazine*, 10.07.2011, Available from http://www.hometoys.com/ezine/09.08/strauss/

Sukumar, S. (1984). Touchscreen Personal Computer Offers Ease of Use and Flexibility. *Hewlett-Packard Journal*, Vol.35, No.8, (August 1984), pp. 4-6

Tsai, T., Chang, H., Wong, A.M., & Wu, T. (2011). Connecting Communities: Designing a Social Media Platform for Older Adults Living in a Senior Village. *Universal Access in Human-Computer Interaction,* Vol. 6766, (July 2011), pp. 224-233

Umemuro, H. (2004). Lowering Elderly Japanese Users' Resistance towards Computers by Using Touchscreen Technology. *Universal Access in the Information Society,* Vol.3, No.3, (July 2004), pp. 276-288

Wagner, N., Hassanein, K., & Head, M. (2010). Computer Use by Older Adults: A Multi-Disciplinary Review. *Computers in Human Behaviour,* Vol.26, No.5, (September 2010), pp. 870-882, ISSN 07475632

Waloszek, G. (2000). Interaction Design Guidelines for Touchscreen Applications, In: Sap Design Guild, 26.02.11, Available from http://www.sapdesignguild.org/resources/tsdesigngl/index.htm

Wright, P., Soroka, A., Belt, S., Pham, D.T., Dimov, S., Roure, D.D., & Petrie, H. (2010). Using Audio to Support Animated Route Information in a Hospital Touch-Screen Kiosk. *Computers in Human Behavior,* Vol.26, No.4, (February 2010), pp. 753-759, ISSN 0747-5632

Yost, K.J., Webster, K., Baker, D.W., Jacobs, E.A., Anderson, A., & Hahn, E.A. (2010). Acceptability of the Talking Touchsceen for Health Literacy Assessment. *Journal of Health Communication,* Vol.15, No.2, (September 2010), pp. 80-92

Ziefle, M. (2010). Information Presentation in Small Screen Devices: The Trade-Off between Visual Density and Menu Foresight. *Applied Ergonomics,* Vol. 40, No.6, (October 2010), pp. 719-730

Part 3

Mobility and New Technologies

6

Universal Design or Modular-Based Design Solutions – A Society Concern

Evastina Björk
NHV- Nordic School of Public Health
Sweden

1. Introduction

Universal Design (UD) is a concept with the aim of promoting the development of products or environments that can be used effectively by everybody without adaptation or stigmatization (Mace, 1985). Modular-based solutions on the other hand can provide the individual with optimum usability as the solutions can be adapted exactly to the needs and requirements of the individual. In this chapter UD product solutions will be discussed in relation to modular product solutions in the perspective of the developing/ manufacturing company and society's request for universal design solutions.

The Committee of Ministers of the European Union - decided to recommend the Governments of the EU member states to accept Universal Design as a philosophy and strategy supporting implementation of full citizenship and independent living for all people, including those with disabilities in the Resolution ResAP (2007)3. According to the resolution; "*Universal design is a strategy which aims to make the design and composition of different environments, products, communication, information technology and services accessible, usable and understandable to as many as possible in an independent and natural manner, preferably without the need for adaptation or specialized solutions.*"

Political ambitions and initiatives such as, for example, the EU Resolution are important and well known prerequisites for implementing a UD perspective into the development of a democratic and integrated society. In addition, activities and advocacy from users/user organizations and nongovernmental organization (NJO´s) also contribute to create a society accessible to "all", and usable by as many people as possible. However, the business sector - the third stakeholder - is involved in the technical and market/sales-oriented development of solutions of various kinds and thereby an important partner when society invest in modern and sustainable public arenas.

Thus, there are three parties forming the BUS triangle - Business, User and Society, which is shown in figure 1 on page 2.

It is in the interest of individual users and the whole of society to have products and environments designed so that they can be used by as many people as possible without having special solutions for every single deflection from the existing norm. Previous studies have identified barriers for increased uptake of UD in companies such as government regulation, training, market data, consumer demand, technical complexity but also lack of knowledge interest and techniques. A study performed in Great Britain concluded that only one third of one hundred companies investigated were aware of the term Universal Design (Goodman et al., 2006). There is a misconception that design for universal accessibility means designing for the elderly and disabled. It has been shown that increased usability and accessibility for older and disabled people benefits users in general because where some are excluded from using a product or service many more are likely to find it difficult or frustrating to use. UD does not eliminate the need for assistive technology (AT). People with disabilities (permanent or occasional) will continue to need AT solutions such as communication aids, visual aids, wheelchairs, orthoses and adapted toys in order to interact more fully with their environment. AT will also be required when UD solutions are lacking due to cost or difficulties in creating good solutions. However, building accessibility into new technologies and curricular materials as they are developed will help to ensure maximal inclusion into the full array of opportunities that are available to all people.

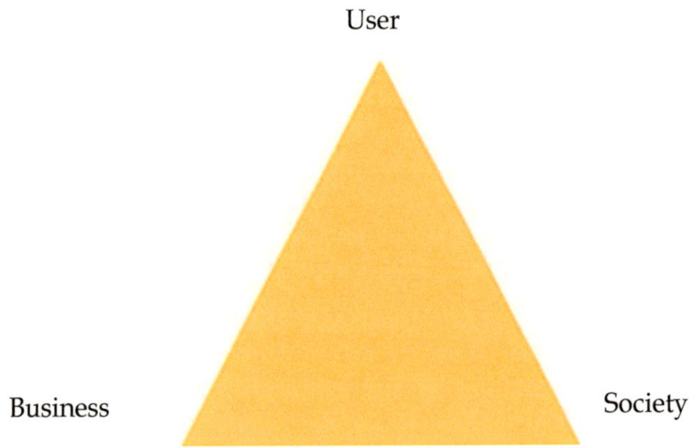

Fig. 1. The BUS triangle illustrates three important stakeholders for the creation of UD solutions (Ottosson, 2004).

By incorporating an attitude of designing products and environments that are universal in use, companies can benefit greatly as markets are expanded and it allows them to identify niche market opportunities. To increase the interest for UD in companies, knowledge and techniques must however be invested in what is in the interest of society due to both legislative and financial reasons.

Disability is not a simple consequence of an individual's impaired capability, but results from a failure to take proper account of the needs, capabilities and preferences of all potential users when designing products, services and facilities aimed towards the public. Elderly people are becoming a demanding wealthy group of customers who want to participate in society, use facilities and services, and who also demand good usability and pleasure in the products they buy. People who suffer from some kind of disability have the same requirements as the population as a whole for accessible environments, usable public transport systems and good usability in products and systems. Legislations, regulations and changing attitudes around the world are generating increasing pressure for a more usable and accessible design. The Norwegian Action Plan for universal design and increased accessibility 2009-2013 is one of several examples (www.regjeringen.no).

The Universal Design concept does not just focus on the user's physical abilities. More attention is on cognitive and communicative abilities, which are sometimes complicated and demanding for designers/product developers to handle. These abilities not readily visible, are difficult to grasp without knowledge within the areas of psychology and sociology and additionally also requires observation of behaviour in performing activities in certain environments over time in order to comprehend. There is more to functionality and task performance than bodily access, and in order to be able to design products and environments supporting behaviour, information on person/environment fit is not enough.

2. Aim

The aim of this chapter is to discuss UD product solutions versus modular product solutions in the perspective of needed technical, financial and other resources for the developing companies. The different conditions will be discussed from the request of society for more universal solutions

3. Modular product design

The basic idea underlying modular design is to organize a complex system (such as a large program, an electronic circuit, or a mechanical device) as a set of distinct components that can be developed independently and then plugged together. In systems engineering, modular design — or "modularity in design" — is an approach that subdivides a system into smaller parts (modules) that can be independently created and then used in different systems to drive multiple functionalities. In a production context modular design in addition to reduction in cost (due to lesser customization, and less learning time) also offers flexibility in design. Modular design is an attempt to combine the advantages of standardization (high volume normally equals low manufacturing costs) with those of customization. Another aspect with modularity is the possibility of adding a new solution by merely plugging in a new module. Computers use modularity to overcome changing customer demands and to make the manufacturing process more adaptive to change. A downside to modularity is that modular systems have a tendency to expand in number of modules which to the customer and user can be at the expense of good usability and for the manufacturer a larger number of parts meaning logistic expansion.

Modular design in an assistive technology context has traditionally been a way to handle the individual needs and demands of people with disabilities. To be able to supply people with individually designed products new modules have been added to already existing product solutions. This has been an unfavourable situation for many children with reduced body function, receiving a smaller copy of equipment aimed and developed for adults. Fortunately today, knowledge about children's needs for good design in assistive devices has resulted in a much better situation. There are however still products on the market that are based on the same idea with the argument that it satisfies the individual need; an argument which is sometimes true but at the same time an argument for not considering a new design solution that perhaps covers the needs of several more users. However, one outstanding argument for modular-based design is the flexibility to change parts in the solution when broken or when user needs change.

From a market perspective it can be argued that a modular-based solution can be more difficult to communicate as it consists of several parts (if not pre-mounted by the manufacturer). As the amount of time companies have to communicate a message to customers is decreasing the message must be easy and intuitive in order to be successful. Simple illustrations can replace long and often difficult written explanations used in user manuals and product instructions and are beneficial as they can be communicated in several markets in multiple countries. Web-based customer/user information is of course the most global and economically most efficient way of presenting information; however, not all households own computers and have an Internet connection. An additional factor is also that the personal computer which was initially considered as a new possibility for accessing information, in fact needed adaptations to be accessible to most users and often people with disabilities have to "wait for the technology"(Emiliani 2009). Supplying companies argue that modular-based products are fully adaptable to individual needs and abilities but the rehabilitation engineers are often forced to rebuild in order to achieve the individual support required. Technical details also in today's solutions are sometimes dimensioned without the important holistic design perspective, e.g. screws, nuts are too big to fit into the product design, there is a lack of user-friendliness to the human hand when a change needs to be accomplished, or a different materials change in colour with time and thereby making the assistive device look old-fashioned. The saying "the devil is in the detail" is indeed true in the assistive technology field as people with disabilities have a reduced ability to adapt themselves to poor product design.

Products communicate with its users through different channels. Interest in the perceptual and image values also in the field of assistive technology products can be seen as a sign of product designers having become aware of that both primary and secondary users of assistive technology constitute demanding and often well-educated customers. It is an expanding, growing market where users also become buyers to a greater extent than previously, which increases the interest of companies for additional product values.

The personal perception of a product is affected by function, perceptual values such as colour, weight, design etc. but also by the image value with its identification attributes. Assistive technology devices have for a long time focused on the technical functions, underestimating other product values.

- *Functional product values* are dependent on the technical solutions often hidden inside the product. When the user can perform the intended activity for which the technical solution was developed, it is a good functionality.
- *Perceptional/sensorial product values* are based on what we experience with our senses (sight/hearing/taste/touch/smell) from outside and/or in contact with a product. The product name's semantics are an important contributor to these values.
- *Image values* are based on the "feeling" the user gets of the product. Brand names, patents, the image given on web pages, stories and the expressed experiences of the product by other users (Ottosson 2004).

4. Universal design

The one running theme in the demographic context is that Universal Design not only provides a framework for action but is an approach that values and celebrates human diversity. Further, as a product of social policy Universal Design can restore equity and enhance citizenship. This can be called the politics of sustainability and civic rights.

4.1 Universal design principles

It can be stated that establishment of the seven Universal Design principles which were initiated by R. Mace and his team at North Carolina State University in beginning of the 1980 was a step towards a human perspective on how products and environments should be designed to be usable and understandable by as many as possible (Connel et al., 1997). The seven principles listed below, (table 1) with the guidelines added to it, gives a slight view of what problems designers and product developers might have had when trying to transform the principles to technical terms, specifications and measures.

Principle	Explanation
1. Equitable in use	The design is useful and marketable to people with diverse abilities *Guidelines* a) Provide the same means of use for all users: identical whenever possible; equivalent when not b) Avoid segregating or stigmatizing any users c) Provisions for privacy, security, and safety should be equally available to all users d) Make the design appealing to all users
2. Flexibility in use	The design accommodates a wide range of individual preferences and abilities *Guidelines* a) Provide choice in methods of use b) Accommodate right- or left-handed access and use c) Facilitate the user's accuracy and precision d) Provide adaptability to the user's pace
3. Simple and intuitive use	Use of the design is easy to understand, regardless of the user's experience, knowledge, language skills, or current concentration level *Guidelines* a) Eliminate unnecessary complexity b) Be consistent with user expectations and intuition c) Accommodate a wide range of literacy and language skills d) Arrange information consistent with its importance e) Provide effective prompting and feedback during and after task completion
4. Perceptible information	The design communicates necessary information effectively to the user, regardless of ambient conditions or the user's sensory abilities *Guidelines* a) Use different modes (pictorial, verbal, tactile) for redundant presentation of essential information b) Provide adequate contrast between essential information and its surroundings c) Maximize "legibility" of essential information

	d) Differentiate elements in ways that can be described (i.e., make it easy to give instructions or directions) e) Provide compatibility with a variety of techniques or devices used by people with sensory limitations
5. Tolerance for error	The design minimizes hazards and the adverse consequences of accidental or unintended actions *Guidelines* a) Arrange elements to minimize hazards and errors: most used elements, most accessible; hazardous elements eliminated, isolated, or shielded b) Provide warnings of hazards and errors c) Provide fail-safe features d) Discourage unconscious action in tasks that require vigilance
6. Low physical effort	The design can be used efficiently and comfortably and with a minimum of fatigue *Guidelines* a) Allow user to maintain a neutral body position b) Use reasonable operating forces c) Minimize repetitive actions d) Minimize sustained physical effort
7. Size/space approach/use	Appropriate size and space is provided for approach, reach, manipulation, regardless of user's body size, posture, or mobility *Guidelines* a) Provide a clear line of sight to important elements for any seated or standing user b) Make reach to all components comfortable for any seated or standing user c) Accommodate variations in hand and grip size d) Provide adequate space for the use of assistive devices or personal assistance

Table 1. The seven Universal Design Principles (Story et al., 2001).

The seven principles define the degree of fit between individuals or groups and their environments, but they also refer to the attributes of products and environments that are perceived to support or impede human activity. They also imply the objective minimizing the adverse effects environments may have on their users such as stress, distraction, inefficiency and sickness. However, the principles require perspective and reflection. Some have criticized their orientation toward products (e.g. Paulson et al., 2005), and others have criticized them as vague, incomplete, and difficult to understand (e.g. Steinfeld, 2002). Although little re-evaluation, reconsideration, or questioning of the principles has occurred since their introduction in 1997, Duncan (2007) suggested adding new principles that relate to affordability and sustainability. He also requested guidance in weighting the principles. Nevertheless, the idea behind the principles remains - to create products and applications that can be used by all customers, independent of their age or physical and mental conditions.

Several studies have shown different barriers and drivers for UD. For instance, a US telephone interview was conducted on 26 consumer product manufacturers and a similar survey on 307 Japanese companies in five different industrial categories (Helen Hamlyn Research Centre, 2000) pointed at government regulations, training, market data, consumer demand, technical complexity and the lack of interest, knowledge and techniques as main barriers.

Goodman (2006) has identified some main causes for the companies lack in acceptance:

- Lack of knowledge
- Lack of business case
- Lack of time and budget

The misconception, that designing for universal usage means only designing for elderly and disabled (Keates et al., 2000) has also been noticed. A UK survey of 29 design professionals stated that "Design for all" was widely known and understood but not widely practiced within the design community (Simms, 2003). Reasons given included lack of time, client backing, money and awareness of the possible market.

Bellerby and Davis (2003) interviewed six product developers and market specialists. They suggested that standards and guidelines could be important drivers but were mostly not presented in an appropriate format. Dong et al (2004) found in a survey of 38 manufacturers in small- and middle-sized companies in the UK that key barriers were based on assumptions such as UD is more expensive, and that there are practical and implementation difficulties.

Goodman et al (2006) presented other drivers for UD; demographic, consumer trends, social responsibility and brand enhancement. Key consumer benefits were increasing customer satisfaction and producing innovation and differentiation.

Most manufacturing companies are using product development models in their design work which has an impact on their ability to adapt to UD principles. Many manufacturers require a UD concept which can be integrated into the Product Development (PD) models they use. Additionally, they require that it can be evaluated towards ordinary quality systems and standards used, otherwise it becomes an additional activity that most manufacturers do not know about or lack knowledge about how to handle or benefit from, or even neglect based on economic or organizational reasons. Blessing (2003) argues that "… *most product development models are static description and to identify whether a method or a tool indeed contributes to success is far more difficult and the results are not easy to generalize. The success of a method or tool depends on the context in which it is being used. This context is different for every design process, because every design project is unique.*"

5. Prerequisites for integration of UD perspective in product development

From reading the literature, scientific publications and based on the author's own experience (Björk, 2003), four main prerequisites that need to be integrated into the PD methodology can be indentified in order to qualify it to act as a guide for a design process applying a Universal Design perspective. The four prerequisites identified are presented and discussed below.

5.1 User intervention with user trials

When referring to user involvement in the implementation of functionality and usability aspects in product design, many authors have reported positive results. E.g. McClelland stated as early as 1995 that user trials are the most valuable source of information about a product's performance. In 1998 Eric von Hippeln discovered that many products and services were developed by users, who then successfully transferred their products to manufacturing in their own enterprises or in other companies. When individual users face problems that the majority of user's and customers do not, they have no choice but to develop their own modifications to existing products, or entirely new products, to solve their issues. Often, user innovators will share their ideas with manufacturers in hopes of having them produce the product. In 1986 von Hippeln introduced the *lead user* concept and argued for the benefits of involving lead users into the development process. He stated that lead users are familiar with actual contexts and have a pre-understanding for special environments which makes them better qualified to identify new products and new solutions (von Hippeln, 2005).

Who then are the users? What kind of different users do we have to consider to cover the range of today's and tomorrow's users? Normally "the user" is seen as just one of the product- or environmental stakeholders but - except from end-users - several others are involved in the usage, such as clients, producers, owners, and decision makers (Nelson & Stolterman, 2003).

However, the individual user perspective has sometimes been shown to be too unilateral and primarily reflecting the person's own situation (Jensen, 2001). The process of user-centered design described at an early stage by Buurman (1997) also argues for user involvement and to assess the performance of users when using products (Jordan et al., 1998; Björk, 2003). Through observation of the user and participation in the usage situation product developers/designers can through their own experiences, knowledge and reflections (the famous reflection in action described by Schön, 1983) achieve a holistic understanding for the actual situation. Studying usage in a real environment by participating as an insider (Coghlan, 2001) allows the product developer/designer access to a great amount of information that is of another character than that which can be obtained via common data collection methods (Ottosson & Björk, 2002). By using all human senses, information input is increased and a holistic view of the situation can be attained.

Improved product usability for those with reduced capability and for disabled people also makes life easier for fully capable individuals (Fig 2). By increasing the uptake of Universal design in companies' development philosophy, the need for assistive devices used by people with disabilities will decrease, which in turn is valuable for all in the perspective of democracy, human rights and equity. One early example is the bus manufacturer Optare which teamed up with the Royal College of Art in the UK to create more interesting bus interiors that can also cater for less able persons (Shahmanesh, 2003). Another company, Gowrings Mobility, is one of the UK´s leading manufacturers and suppliers of wheelchair passenger vehicles. The managing director of the company explained that a design engineer should have a more realistic understanding of who is actually using the car. Where "no cost choices" can be made at the design stage, this can benefit a whole range of drivers and passengers of all ages, abilities and sizes (Shahmanesh, 2003).

Another valuable tool for obtaining user involvement grasping information is focus groups, but they are rarely used in isolation. Marketing researchers employ a variety of tools, including one-on-one interviews, written surveys and polling to track consumer opinion. Used together with all of the above, a focus group is an integral part of gauging public perceptions. Focus groups have some obvious benefits: The product developer can interact with the participants, pose follow-up questions or ask questions that probe more deeply. Results can be easier to understand than complicated statistical data and the product developer can get information from non-verbal responses, such as facial expressions or body language.Information is provided more quickly than if people were interviewed separately.

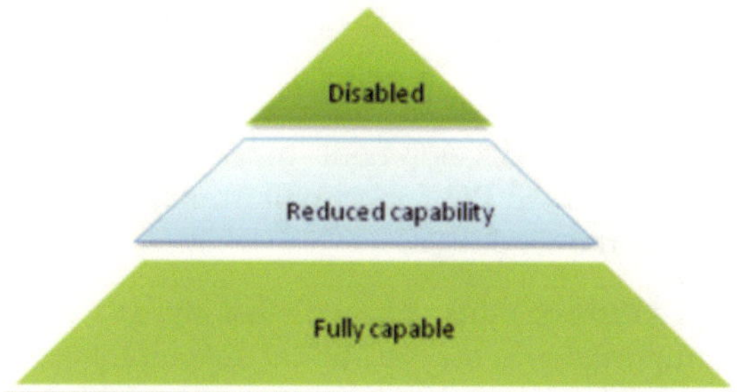

Fig. 2. The user pyramid (Ginnerup, 2010).

While all of these are valid points and give more information than a survey or questionnaire, they do not always give as much as is needed to succeed. The small sample size means the groups might not be a good representation of the larger population and the group discussions can be difficult to steer and control, so time can be lost to irrelevant topics. Additionally respondents can feel peer pressure to give similar answers to the moderator's questions, the moderator's skill in phrasing questions along with the setting can affect responses and skew results (Edmonds, 2000).

5.2 Focus on user's desires

It is meaningful to distinguish between a present user need, a want or a more distant desire. Incremental innovations are often based on satisfying a present *need*. Radical innovations on the other hand are often based on satisfying a *desire* – or *wish* as it has originally been termed (Ottosson, 2006). The conditions for *want*- and *desire*-based Product Development methodology differ much from *need*-based PD for which the well-known PD models were initially designed. Some differences between the three PD driving forces mentioned are shown in Table 2.

Driving PD force	**Characteristics**	**PD Target**	**Planning**	**Stable conditions**	**Unstable conditions**
Need	Knowledge and solutions exist to re-use for an existing need	Fixed	Fulfil plan	Yes	No
Want	Knowledge and solutions are incomplete to solve a new want	Moving	Adopt to the situation	Partly	Partly
Desire (Wish)	Important knowledge and solutions do not exist	Vision	Create, make and test	No	Yes

Table 2. Three types of backgrounds for product development causing different circumstances for PD work (based on Holmdahl 2007).

Generally, a need-based PD project has stable conditions, a want-based PD project experience more partly unstable conditions while a desire-driven project represents totally unstable conditions. Two philosophically different views exist on how to best perform need-based PD development, leading to a categorization of PD methods as either classic or dynamic depending on their ability to handle stable/unstable conditions.

For a company performing PD projects based on a market *need,* the time factor is crucial from a market perspective as the need/problem already exists and the risk for competitive solutions to occur is great. The price is the second most important variable as many similar solutions can appear on the market, meaning a price competition. In turn that means a demand for low PD and production costs as well as effective logistics. User intervention is a valuable resource in need-based development projects and focus group can be one alternative to consider. A need-based PD project will from a company perspective therefore benefit from modular-based product design.

For the development of products or other solutions that are based on a *want,* the time factor does not have the same importance as for *need*-based development. The long-term planning in these projects is not possible to maintain as so many variables are unknown when the project starts; planning can only be successful for short periods of time. The market price is not at all an issue initially in the desire-driven development projects (often innovation projects). Especially lead users (von Hippeln, 2005) can initially make important contributions for products or solutions that are based on a *desire,* where the time and price factors are less important. From a company perspective, a desire-based PD project would likely benefit from a UD solution. To find out *wants* for a near future - or *a desire* for a more distant future - lead users and dialogues with professionals in certain fields can be of good value. End-users, rather than manufacturers, are responsible for a large amount of new innovation (von Hippeln, 2005).

To be able to create solutions which are inventive and attractive and which go beyond today's user needs and functionality the focus should be on users' desires. Edefors (2004) discussed the problems with focusing on the actual local user and argued for a wider perspective. The presumptive users of tomorrow are interesting to investigate as they have the arguments and motives for **not using** an actual product or environment today. Their motives for being no users put demands for new inventive solutions. This approach is similar to Jordan (2002), he argues for fitting products to people in a holistic manner where the relationship between people and products depends on more than just usability, it is about perceptual and image values. People have hopes, fears, dreams, aspirations testes which influence their choice and experience of product and environments (Björk, 2003).

5.3 The product development methodology should be able to deal with complexity

Manufacturing industries are under tremendous pressure to reduce cost and time-to-market and yet offer a large variety of products. Consequently, the companies are compelled to operate at the lowest profit margin and shift manufacturing operations to developing countries for cost savings. Current PD methodology is not adequately upgraded nor equipped with efficient design tools and techniques to meet the challenges of the Universal Design concept on a global market, as mentioned earlier.

Complex systems consist of a large number of dynamically (and usually non-linearly) interacting non-decomposable elements (McKelvey, 2004). Because of high interconnectivity between elements, it can often be difficult to associate effect with cause (Holmdahl, 2007).

One is confronted with incredible intricate interacting parts and not relatively easily identifiable chains of cause and effect apparent in linear processes. The world is experienced as being increasingly complex, unordered and non-linear which becomes obvious in innovation projects. The complexity comes from the fact that such projects build on limited number of pre-known solutions.

To be able to manage innovative product development projects which are unique and difficult to plan beforehand, a dynamic and flexible philosophy is needed. Most product development methodologies of today are created with big- or middle-sized firms in mind where certain rules and hierarchies are established and where the company system rules all activities. The Stage Gate system philosophy (Cooper, 1986) is well known but cannot offer flexibility, short planning periods or uncertainty as it is a static method.

The world is changing faster than ever before. Market dynamics increase, changes in fashion with different trends shift rapidly and product life is reduced. If companies do not get their product on the market at the right time, it flops. Laws which regulate the conditions for product design, sales, production and destruction change frequently.

If, in such a situation, one starts product development with a detailed specification and a detailed schedule and sees product development as a matter of delivering specifications and follow the plan, several problems will occur. Things are even worse if a serial/sequential development strategy (baton method) with different phases that traversed sequentially is selected. If the project also slows down by using different gates (toll gates), the situation can go really crazy in a changing market.

A dynamic strategy is required to meet the ongoing changes and no one can in advance know exactly how the product should be wired to best suit a universal market. *Dynamic Product Development (DPD™) method* (Ottosson, 2004) argues that flexibility and easy adaptation to changing circumstances must be built into the system. The organization must be competitive in order to quickly respond to new impulses and new insights. Only an expert-led self-organizing organization has these characteristics. *Dynamic Product Development (DPD™) method* is based on a usability philosophy; good usability is based on knowledge about human performance and of the environment where the performance takes place. To cope with complexity and unstable situations, DPD™ argues for the presence and participation of the Designer/PD developer in the project mentally and physically. By being present in the centre of the development immediate feedback from the development activities can be gained.

Usage involves human behaviour which is strongly linked to context, environment and time and is a good example of a complex relation. The study of usage in actual environments as has been focused on in this chapter is one important tool to find out about what tricky situations there are to be handled.

5.4 Interdisciplinary teams

To be able to effectively deal with Universal Design-based development, a lot of knowledge and experience from separate fields is needed, which is why various professions should be represented in the teams. This prerequisite is almost adopted in all development processes in society today. Team composition has proven to be of extreme importance for outcome.

How one design the team is said to affect the performance 40 times more than coaching a team (Hackman, 2002). There is a risk of achieving too little cohesion in a group of disparate talents, and if there are personality differences, communication within the team is hampered. A heterogeneous group performs better than a homogeneous (Pech, 2001) as most decisions taken are intensively motivated, the group represents a holistic view and different perspectives are represented. Respect for other knowledge than that which you yourself have is important in all forms of team work. Team leaders' job is often a balancing act between achieving goals, deadlines and cost limits, which is why important communication within the team and with other stakeholders might sometimes be less than required.

Several of the projects initiated by municipalities or regions and relating to the design or equipment in the public domain are purchased through so-called public procurement. Companies are asked to give a quote for the product or services requested based on a list of demands put together by the purchaser. Low price is the most valued variable when the offers are examined and variables like accessibility, usability, etc. are not addressed at all as knowledge is lacking among the purchasers on how those demands could be addressed in the list of demands. Interdisciplinary teams are needed at different levels in the system to prevent that disabled and other groups become excluded and discriminated, ultimately prevented from using the public domains.

Understanding user requirements on a holistic basis requires a focus and attention to the roles that products play in people's lives. Ethnographic methods may be particularly useful in this context as their use may give rich insight into the roles that people have in different situations in life. Such methods tend to be qualitative in nature.

6. Some examples of UD visions in companies today

6.1 Toshiba

Toshiba America, Inc. (TAI) is the holding company for one of the leading groups of high technology companies, with a combined total of approximately 8,000 employees in the U.S. Together, the U.S.-based companies under TAI's umbrella manufacture and market represent a widely diversified range of modern electronics, each conducting research and development, manufacturing, sales and service in its field of expertise.

Toshiba Group is collaborating with internal and external specialists on product development in various fields, including home appliances, housing facilities, information equipment, and public facilities. By applying a human-centered design process emphasizing the users' perspectives and incorporating customer requirements, the Toshiba Group believes they contribute to realization of a society where everyone can live at ease and in comfort regardless of age, gender and abilities"(www.toshiba.com).

On their website an ambition has been formulated which tells about a new way forward "Transforming "can't use" to "able to use", "hard to use" to "easy to use", Toshiba's universal design aims to create products accessible and safe to use for everyone. With the perpetual drive for innovation, Toshiba continues to explore ways to create more convenient and easier to use products which, will meet even greater number of peoples' standard for "want to use"(www.toshiba.com).

The Universal Design (UD) Promotion Working Group (WG) established in 2005 is striving to incorporate the universal design concept in development steps and is promoting dissemination of information on Toshiba's universal design internally and externally. The triangle in fig 3 constitutes of the three main and important parts which Toshiba argue are the most important for creating products accessible and safe to use for everyone. Some of the seven UD principles have been adopted as specially important.

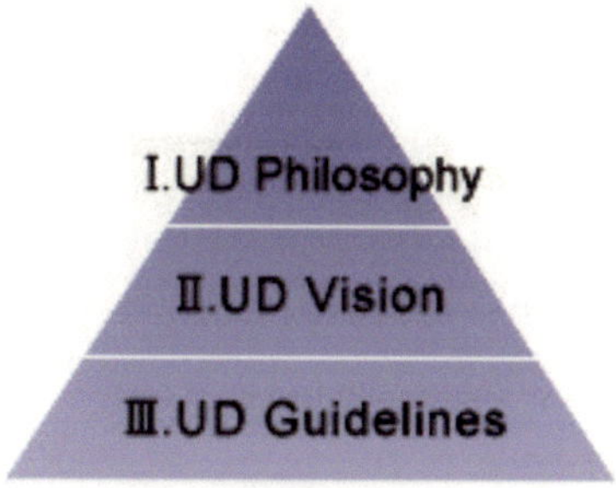

Fig. 3. Toshiba Group Universal Design Guidelines.

- Intuitive use
- Simple use
- Low physical effort
- Equitable use
- Safety and minimal anxiety

6.2 Omron

Omron Corporation was founded in 1933 in Kyoto, Japan, and Omron Healthcare is a subsidiary of that company. Omron is one of the leading distributors of medical, home health care, and wellness products worldwide.

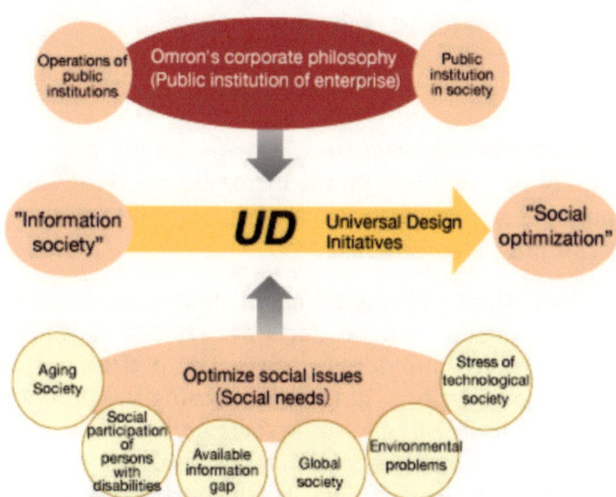

Fig. 4. Omron's idea of Universal Design.

Omron has a vision and a philosophy which is "Sensing tomorrow"(www.omron.com). That vision is from a demographic perspective a very good one and as customer satisfaction is in focus the promotion of Universal design to make products easy to use by a broad range of people it fits in to the companies idea of Universal design.

6.3 The Careva systems AB

Careva Systems AB is a small-sized Swedish enterprise started in 1998. It specializes in positioning equipment for safe and comfortable transportation of persons with disabilities in vehicles cars, vans and buses. The vision behind the development of the company's existing product range is that "everyone should have the right to travel in vehicles in a safe and comfortable way. To be able to realize the vision two product lines have been developed" (www.careva.se). Careva's idea of UD is that all kinds of users should be able to use all kinds of transport systems

Fig. 5. The idea behind the Company´s product line.

1. *Careva belt - a modular based system* for optimization of the individuals requirements, differently designed parts can be mounted together to meet the personal requirements of the individual, a traditional assistive technology device.
2. *Crossit belt - a universal design solution* meaning that the same solution can be used by most users who requires a positioning support in vehicles without individual adaptation and additional items.

The company´s strategy to offer both modular and UD solutions is interesting. As the modular based system were introduced to market several year ahead of the UD solution which was introduced to market less than a year ago. Perhaps the company discovered the new trends and a market request for UD solutions.

6.4 Society concern

In Figure 6 below the relation between the "Private room" and the "Public room" is illustrated with the aim of putting focus on the increasing need for intervention but also on the difference in requirements for how environments and products are designed dependent on the individuals performing activities there.

First, in the "Private room" (here understood as the home), the focus is on the individuals living there and the environment and products are created towards their requirements (often in the family context), an individual design. The persons act as users but also as customers, meaning they use their own judgment choosing the products they buy and the environment becomes a result of their choices and creativity. However, today many of the people who live with a mental or physical disability reside in group homes, institutions, nursing homes or at home with their parents. This means that somebody else is in control and setting the rules. Even if a person may be in need of assistance it is also important that he or she have a measure of autonomy. Most people experience the advantage of living in their own home as it means one can be in control.

The private room must fit on a detail level to promote the independence and empowerment of the individual without being an institutional setting which could have negative effects for other family members. Home modification and Assistive Technology make it safer and easier for people with disabilities to live independently Assistive technology is in this context defined as; technology used by individuals with disabilities in order to perform functions that might otherwise be difficult or impossible. If Universal Design solutions could enter also into the private room it would facilitate intervention and reduce the number of special solutions in the field of AT.

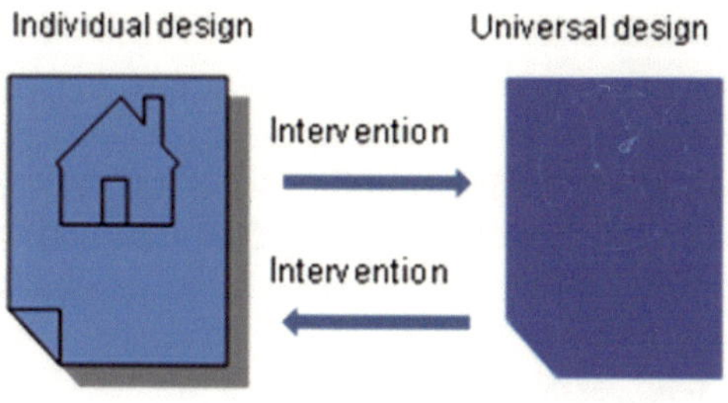

Fig. 6. The relation between the "Private room" and the "Public room".

In the "Public room" on the other hand, solely Universal Design solutions are required to optimize usage by as many people as possible; a true demand for professionals designing buildings, outdoor environments and services. How to design an entrance in a public building? How to design the information system in the public transportation system to be accessible and understandable to as many as possible? What is to consider when purchasing benches for use in a public park? How to create a playground with accessibility for children with different abilities? Professionals compete to offer "the best solution" for the city or the municipality after considering all the demands that have been sat up by the purchasing representatives. The competence and experience among the purchasing representatives is of outmost importance as the public procurement establishes the rules for what solution should be accepted. Unfortunately the purchasers are not familiar with the Universal Design concept but consider cost as the most important variable. Knowledge about UD in Public procurement processes have a huge impact on how fast the UD concept can be accepted and implemented in public arenas in society.

Laws and regulations have been established in several countries to safeguard to democratic rights and to prevent segregation due to ability, age, gender or ethnicity. They are meaningless, however, unless wedded to policies and practices that challenge the realities of property development and design dynamics. Economic and cultural rationales and values drive these realities and, in doing so the needs of diverse users of the built environment are often overlooked (Imre & Hall, 2001).

One area obviously not prioritized at design/engineering schools is learning about the Universal Design concept and how to perform real UD solutions. Companies and consultancy firms need guidance to be able to fulfil the intentions of an inclusive society. Except from implementing laws and regulations, society should take responsibility for the change towards UD solutions in the public room getting the right prerequisites otherwise it will be a flop.

The intervention becomes more intense the number of elderly become more mobile and require public services. Physical accessibility has also improved during recent years, which offers new opportunities for groups of users who were previously locked out. The private room has furthermore become a working place where home services of different kinds need to be carried out and where demands for a good and safe working environment for staff is present. All aspects together make Universal Design a socioeconomic factor, a human/quality of life factor as well as a market/business factor.

6.5 The demographics

The number of individuals with some form of disability has, in Europe, been estimated to be between 12 and 15% of the population, a figure which is increasing due to a growing number of elderly but also the fact that disability statistics tend to count those individuals who are registered as permanently disabled, occasional disabled who would benefit from a more accessible environment or increased usability in products is not included in the figures. The consequence is that disability statistics almost certainly underestimate the number of individuals who experience limitations in activities due to reduced body function.

Demographics require companies to abandon the concept of targeting only young and fully capable customers (Grassman & Reepmeyer, 2008). They need to create new products that are attractive to a broader customer group. Products that follow the principles of Universal Design do not separate but integrate customer groups, and they can also substantially increase a company´s targets.

The 50+ generation turns out to be one of the most attractive target groups. The "new" elderly generation is much more vital and has a higher purchasing power compared to the "older" generation. Research on the economic potential of demographic change is fairly limited. Most studies relate to customer segmentation approaches and defining the needs and special demands of older customers (Grassman & Reepmeyer, 2008). Considering the future, companies have no other choice than developing products independent of customer´s age and ability to be successful. The concept of Universal Design represents a standard, not an exception, and Universal Design intentionally avoids highlighting the users' and the customers' different capabilities.

In order to identify and define processes that allow implementing Universal Design strategies, a closer look at the older customers' specific capabilities and abilities has proven to be a very effective first step. Physical and mental capabilities usually worsen with old age. Many age-related medical conditions lead to the fact that sensory capabilities and velocity-related activities decline. Vision usually starts to deteriorate at an age of around 40. The eye's abilities relating to contrast and colours are usually impaired by then. Starting at around 60 years of age, more severe constraints frequently occur. Hearing impairment usually starts at an age of around 60, the same age where the muscular strength usually starts declining as well. For example, a 60-year-old has on average 15–35% less muscular strength than a 20-year-old (Grassman & Reepmeyer, 2008).

Taking all these factors into account, adding also the changes in mental capability researchers in gerontology have come to the general belief that people beyond their mid-60s face considerable multi-morbidity issues (Grassman & Reepmayer, 2008). However, studies have shown that substantial losses of physical fitness only correlate for people with an average age of over 85. Figure 7 illustrates the increasing need for assistive technology in older age.

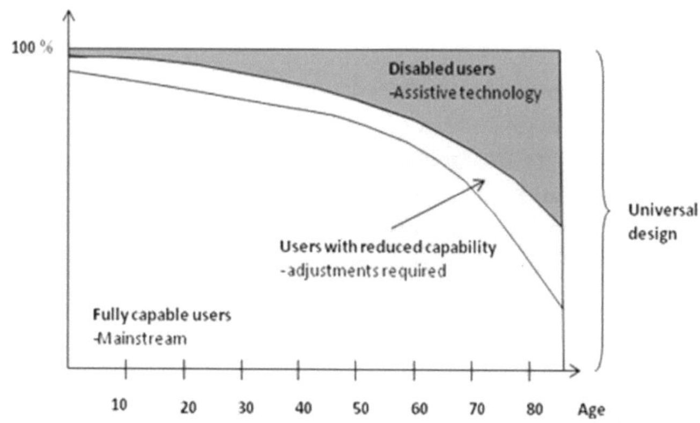

Fig. 7. An illustration of how different groups of users are represented in relation to age (Ottosson, 2009).

Users with severe mobility problems (in black, in Figure 7) require assistive technology, often individually developed solutions. Mainstream products and modular based systems can also be adjusted to fulfil their requirements. Users with reduced capability (in grey, in Figure 7) represent those who require smaller adjustments of mainstream products to maintain independence. Often modular-based systems can be used without adjustments.

Figure 7 also illustrates that below the age of 50, about 80-85% of all users are fully capable and use mainstream technology products. Over the age of 50 the number of users who require assistive devices or adaptations increases. Users with disabilities increase nearly linear from the age of 60 up to 80 years of age.

7. Conclusion

The discussion in this chapter has focused on the prerequisites companies require to be able to be an active part in the BUS triangle for an inclusive society. Without the physical solutions products and systems supplied by the business sector the argumentation becomes more of a vision than a reachable realistic goal. In the argumentation for a Universal Design perspective to the benefit of a democratic and inclusive society, more is required to make the companies invest in UD solutions.

What is too little discussed is the complexity connected to the development of UD solutions and the fact that when innovations and new products are required, modular design is often inappropriate. As time-to-market is longer and project costs are often higher for universally designed products in a short perspective compared to modular systems, there can be limited commercial reasons to invest in universally designed solutions in a short-time perspective (Björk 2010). Some benefits have been argued for companies to invest in UD solutions:

- Can be quickly communicated to customers/users.
- Covers a bigger market.

- Requires less logistics in production and marketing compared to modular-based design products.
- Contributes to savings in marketing.
- Contributes to innovation and future expansion toward a sustainable society.

Of the three main parts comprising the BUS triangle, society and the user clearly benefits from UD concept. However, the business sector, supplying with the physical solutions have to calculate costs with higher margins than for modular-based or mainstream technology products. The time to break-even in these projects is many times longer due to the unstable and complex development circumstances. Nevertheless, within a longer time perspective UD solutions might contribute to success. Consequently, support from society, both in financial terms and in improving competence in industry, is essential to ensure that new methods for product development become known and practised for guiding the creation of UD solutions.

8. References

Bellerby, F.,Davis, G. (2003). *Defining the limits of Inclusive Design,* Proceedings of Include 2003. London Royal College of Art pp 1:00-1:17

Björk, E. (2003). *Insider Action Research Applied on Development of Assistive Products,* PhD Thesis, Otto-von-Guericke University, Magdeburg, Germany.

Björk, E., Ottosson, S. (2007). *Aspects of Consideration in Product Development Research,* Journal of Engineering Design, 18(3): 195–207.

Björk, E. (2010). *Why did it take four times longer to create the Universal Design Solution?,* International Journal of Technology and Disability, 21(4): 159-170.

Blessing, L. (2003). *Future issues in Design Research* in Lindemann, U. (ed) *Human behaviour in Design,* Springer Verlag ISBN 3-540-40632-8, Germany, pp 298-303

Buurman, R.D. (1997). *User-Centered Design of Smart Products,* Ergonomics, 40(10): 1159–1169.

Coghland, D. (2001). *Insider Action Research: Implications for Practising Managers,* Management Learning, Vol. 32, No 1, pp 49-60

Connell, B.R., Jones, M., Mueller, J., Mullick, A., Ostroff, E., Sanford, J. Steinfeld, E., Story, M. & Vanderheiden, G. (1997). *The Principles of Universal Design,* Version 2.0, North Carolina State University, Raleigh, North Carolina. *http://www.design.ncsu.edu/cud/about_ud/udprincipleshtmlformat.html*

Cooper, R. (1986). *Incorporating Human Needs into Assistive Technology Design,* in Gray. David B. et al., *Designing and Using Assistive Technology,* Brookes Publishing Co, Baltimore, Maryland, pp 151-170

Council of Europe (2007). *Achieving full participation through universal design.* Committee of Ministers Resolution ResAP (2007)3.

Dong, H., Keates, S., Clarkson, PJ. (2004) *Inclusive Design in Industry, barriers, drivers and business case.* Proceedings of ERCIM Workshop "User interface for all", Austria

Duncan, R. (2007). *Universal Design Clarification and Development,* Internal report for the Norwegian *Ministry* of Environment, Government of Norway.

Edmonds, H. (2000) *Focus group Research Handbook,* ISBN ISBN-13: 978-0658002489 Mc Graw

Edefors, H. (2004). *Product development and design and other paradoxes,* PhD Thesis, Lund University, Dept. of Design and Art, Sweden.

Emiliano, L. (2009). Perspectives on Acessibility: From Assitive technologies to Universal Access and Design for All,(in the Universal Design Handbook, ed. Stephanidis.C., CRC Press, Taylor & Frances, pp.1:1 - 1:11

Mullick, A., Ostroff, E., Sanford, J. Steinfeld, E., Story, M. & Vanderheiden, G. (1997). *The Principles of Universal Design,* Version 2.0, North Carolina State University, Raleigh, North Carolina.

Ginnerup, S. (2010). *Achieving full participation through Universal Design,* ISBN-13-9789287164742, Ingram International

Goodman, J., Dong, H., Langdon, P. & Clarkson, P-J. (2006). Increasing the uptake of inclusive design in industry, *International Journal of Gerontology,* 6(3).

Grassman, O. & Reepmeyer, G. (2008). Universal design - Innovations for all ages, in Kohlbacher, F & Herstatt, C (eds.), *The Silver Market Phenomenon: Business Opportunities in an Era of Demographic Change,* ISBN 978-3-540-75330-8, Springer,Berlin.

Hackman, J.R. (2002). *Leading teams. Setting the stage for great performances,* Harvard Business School Press, Boston ISBN 1-57851-333-2.

Holmdahl, L. (2007). *Complexity Aspects of Product Development,* PhD thesis, Otto-von-Guericke University, Magdeburg, Germany.

Imrie, R. & Hall, P. (2001). *Inclusive Design: Designing and developing Accessible Environments,* Taylor & Frances, New York

Jensen, Lilly (2001). *User Involvement, in Development and Evaluation of Assistive Technology, Nordic Development Centre for Rehabilitation Technology,* NUH, Tammerfors Finland, Joma Tryck AB

Jordan, P.N. (1998). *Human Factors for Pleasure in Product Use,* Applied Ergonomics,29(18):99, 25.

Jordan, P. (2002). Designing pleasurable Products, Taylor&Frances ISBN-13: 978-0658002489

Keates, S. et al (2000). *Investigating industries attitudes to universal design,* proceedings of Rehabilitation Engineering and Assistive Technology Society of North America, RESNA, Orlando, USA

Mace, R.L. (1985). *Universal Design, Barrier-Free Environments for Everyone,* Designers West, Los Angeles.

McKelvey, B. (2004). Toward a complexity science of entrepreneurship, *Journal of Business Venturing,* 19: 313–341.

Nelson, H. & Stolterman, E. (2003). *The Design Way -- Intentional Change in an Unpredictable World,* Educational Technology Publications, New Jersey.

Ottosson, S. (2004). *Dynamic Product Development* - DPD™, Technovation - the International Journal of Technological Innovation and Entrepreneurship, 24: 179–186.

Ottosson, S. et al (2006). *Research Approaches on Product Development processes,* International Design Conference – Design 2006, Dubrovnik, Croatia

Ottosson, S. (2009). *Frontline Innovation Management – Dynamic Business & Product Development,* Ottosson & Partners AB, Sweden (ISBN 978-91-977947-7-0).

Paulsson, J. (2005). *Universal Design Education Project – Sweden,* School of Architecture, Chalmers University of Technology, Gothenburg, Sweden.

Pech, R. J. (2001). *Reflections: Termites, Group Behavior, and Loss of Innovation*: Conformity Rules!, Journal of Managerial Psychology, 16(7/8): 559.

Prakash Y., Bimal P., Rakesh J. (2007). *Managing product development process complexity and challenges: a state-of-the art review*, Journal of Design Research, 6(4): 487–508.
Schön, D.A. (1983): *The Reflective Practioneer*: How Professionals Think in Action, Basic books, New York
Shahmanesh, A. (2003). *Age Concern*, Automotive Engineer, January 34–36.
Sims, RE. 2003. *Design for all: Methods and Data for Support Designers*. PhD thesis, Loughborough University, Loughborough, United Kingdom
Steinfeld, E. (2002). *Universal Designing*, in Christoffersen, J. (ed.): *Universal Design*, ISBN 82-90122-05-5, pp. 165–189, PrePress ,Haslum Grafiske, Norway
Story, M.F., Mueller, J. & Mace. R. (2001). *The Universal Design File, Designing for People of all ages and Abilities*. North Carolina State University, Raleigh, North Carolina.
Barne-, likestillings- og inkluderingsdepartementet [Ministry of Children, Equality and Social Inclusion]. (2009). Norge universelt utformet innen 2025. Regjeringens handlingsplan for universell utforming og økt tilgjengelighet, Government of Norway. URL: *www.regjeringen.no.*
Wijk, M. (1997). *Differences we share*, Faculty of Architecture, Delft University of Technology, Netherlands.
von Hippel, E. (2005). *Democratizing Innovation,* The MIT Press, Cambridge, Massachusetts, London, England.
von Hippel, E. (1998). *The Sources of Innovation,* ISBN 0-19-509422-0, Oxford University Press, New York
von Hippeln , E. (1986). "Lead Users: A Source of Novel Product Concepts", *Management Science* 32(7): 791–806, http://www.jstor.org/stable/2631761
www.toshiba.com, http://www.toshiba.co.jp/design/pr/ud/intro.htm
www.omron.com, http://www.omron.com/media/press/2008/03/h0325.html
www.careva.se,http://www.google.com/search?hl=sv&q=careva+systems+Universal+design

7

Proposal for a New Development Methodology for Assistive Technology Based on a Psychological Model of Elderly People

Misato Nihei and Masakatsu G. Fujie
The University of Tokyo, Waseda University
Japan

1. Introduction

Developed countries are all witnessing an increase in the aging of their populations and Japan is no exception to this. In fact, Japan is a pioneer in this respect, and as of 2011 the nation's elderly population stood at 29,440,000, or 23.1% of the total population (The Cabinet Office in Japan, 2011). Social aging has thus triggered research into geriatric assistive technology (AT) as well as a supportive societal infrastructure catering to the elderly and to persons with disabilities. The AT industry in Japan, which has great significance for social and industrial users, has been referred to as a growth industry since legislation on AT was enacted in 1992 (M. Watanabe et al., 2006). Some modifications to the social system and care insurance system were enacted in 2000, and the market has been steadily expanding and currently it is said to have reached maturity (K. Masahiro, 2005). In the past decade, emphasis has been given to developing an AT infrastructure that can readily facilitate the adoption of the advances being made in standardizing AT and to further improve the quality of this service in addition to enhancing the safety and operability of the system (S. Hashimoto,.2007).

Unfortunately, the gap between AT developers and users in understanding the needs involved is a wide one that is just beginning to be addressed, e.g., the development and standardization of databases focusing on the individual physical and psychological profiles of the elderly (Body function Database of elderly persons, 1993); the effectiveness of the systems (Y. Shinomiya, 2001); and the emotional and mental effects of the AT on users in terms of such matters as physical function assistance versus mobility. In this study, we analyzed the AT-user relationship and presented the results in a form that AT developers can understand and apply. Taking an inductive approach, we studied changes in the mental states of AT users and looked at device-oriented lifestyles. To build a useful psychological concept design model, we used psychological analysis methods widely employed in medical science, psychology and sociology. As the focus of our study, we chose the wheelchair as a representative example of AT as an example that is both widely used and widely studied (R. Morales, et al., 2006; Q. Zeng, et al, 2006; J.L. Murray et al., 2003).

The use of ATs can improve the physical and social functioning of persons with disabilities and elderly people (J. Jutai, 1999). When developing ATs, engineers need to take an inclusive approach that considers user participation and user needs (T. Inoue et.al, 2002). The development process should also emphasize conceptual design, based on the physical and living situations of targeted users. Nonetheless, some persons inevitably feel a conflict with regard to the use of ATs. We hypothesize that, for a number of elderly people, the use of ATs is associated with a degree of psychological resistance (Figure 1). According to some reports, such psychological resistance arises and bears a social stigma (B. Louse et al., 2002), when ATs are selected and introduced by therapists (DJ. Baker et al., 2004). On the other hand, there are certain psychological effects produced as a result of using ATs. Consequently, in this study, we propose a new and effective method of developing ATs. In the proposed method, psychological factors associated with the use of ATs are carefully examined before formulating the concept for a new AT.

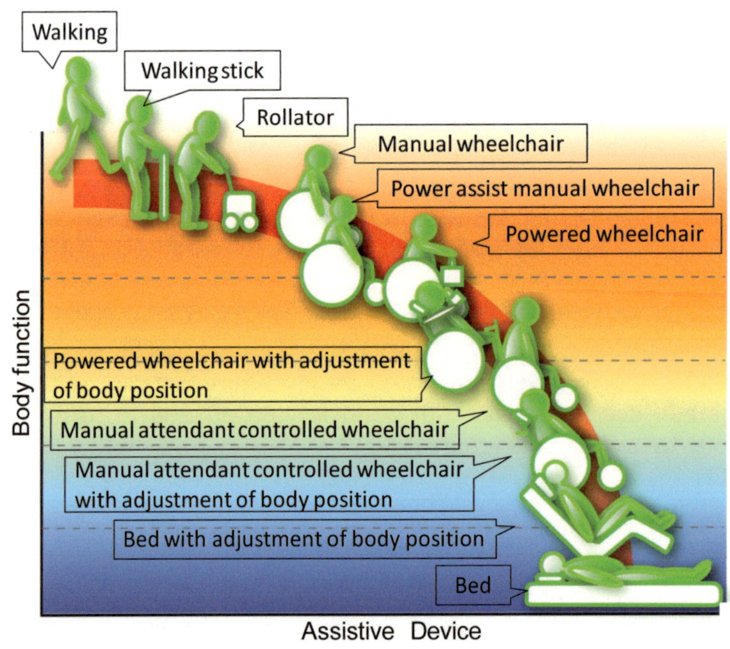

Fig. 1. Relationship between body function and ATs.

In this study, complicated psychological situations are taken into account at the conceptual design stage. The process of developing ATs comprises the following phases:

1. To clarify the process by which psychological conflict with respect to the use of AT is generated, the qualitative research approach was selected.
2. A psychological conceptual model of elderly persons using ATs is developed.
3. The concept of a mobility aid is developed based on the psychological conceptual model and discussions with users.
4. A prototype of a new mobility aid is constructed.

2. The position of a quantitative research approach in design

2.1 The effect of devices on human psychology

In recent years, there have been discussions on what benefits have been seen by using ATs. In particular, it has been made clear in some studies in the practical clinical field that ATs as well as the decrease in daily life activities, and the physical burden on caregivers and that means that the effects on the total evaluation, including psychological. Inoue et al., based on case studies of person with severe disability, state that AT (a powered wheelchair as a case study) has a potential to improve physical health, activity in daily life, social life and, indirectly, spiritual life by using the component QOL (quality of life) model of Lorentsen (T. inoue et al., 2002).

On the other hand, there are several reports of AT not being used in the clinical field. The acceptance of a disability is considered to be one of the causes of this AT non-use. Rosalind et al. describe how physiological and social care became necessary when therapists started to introduce ATs to users in the 1980s (H. Rosalind et al., 1998). In the literature, Krueger's disability acceptance model explains that shock, expectations of recovery, confusion, efforts to adapt, adaptation have a lot to do with why and how users accept the use of a wheelchair.

2.2 ATs for persons with disabilities

ATs help people live their lives in general. So what impression does a wheelchair user make on another person? In a study that examined the awareness of ATs for persons with disabilities, positive aspects such as "(AT) raise awareness of independence," "Give a sense of safety," (M. Roeland, et al., 2002) and "freedom (D.J. Baker, et al., 2004)" were described. On the other hand, negative aspects such as "limits activity (D. J. Baker, et al., 2004)" have been reported. In interviews with persons with disabilities conducted by Lupton et al., positive aspects such as "it is possible to clear various obstacles" were reported (D. Lupton, et al., 2000). On the other hand, negative aspects such as "(assistive devices) to the perceived self-barriers," "really feel that people with disabilities," "difference" and "defect"—indicating a social stigma—have been pointed out as a potential downside to the use of AT.

2.3 Different between the elderly and persons with disability

Based on these reports on persons with disabilities, the particularity and issues were clarified and policy recommendations and medical interventions were proposed. However, in cases that focus on the elderly, the difference in characteristics between the elderly and persons with disability were not clarified. From the few reports related to AT research for elderly person without disabilities, there are statements of opinion or concept regarding the meaning of AT for the individual user. However, these reports did not refer to the development of AT. Therefore, the aim of this study is to investigate the needs of the elderly with respect to AT and also to research the factors and processes behind non-use, and whether it is the same as for persons with disabilities.

3. Assistance for mobility activities (M. Nihei, et al., 2008)

3.1 Research questions regarding assistance for mobility activities

We reviewed the literature and classified prior research into: effects of AT use on QOL, effective AT applications, problems arising in clinical settings, user response to AT applications in a clinical setting, and the meaning of AT for users and user awareness of AT. In addition, we considered the following in the light of prior research:

- Psychological considerations are related to AT non-use
- It remains unclear whether the disability acceptance model is applicable to the elderly
- Study results must be visualized in relation to AT development

Specifically, we sought to define:

- Factors leading the elderly to like or dislike wheelchairs
- How these factors are related to nonuse of wheelchairs
- How the elderly view wheelchairs when using them
- Factors for differences in views of wheelchairs among elderly users and nonusers

3.2 In-depth interviews

This paper investigates how elderly people feel about ATs. In this study, we examine two types of elderly people: those who have never used mobility aids; and those who are using (or have used) mobility aids in their daily lives (Figure 2). A semi-structured interview approach was selected for analysis. First, profiles of the subjects were constructed through interviews, and their living conditions and physical situation were recorded based on their responses regarding environmental and personal factors. Next, questions concerning 13 items were designed and categorized into: living space, mobility walking experience of wheelchair use, and capability of walking categories. The primary factors were extracted via a recording process, based on a transcript of language data. Consequently, a number of factors were found to impact the subjects' impression of mobility aid use, including the image of mobility aid uses, the impression of mobility aids, comparison with other persons, a new understanding of the walking function, acceptance of their situation, and emphasis of the advantages of mobility aids (shown in Table 1).

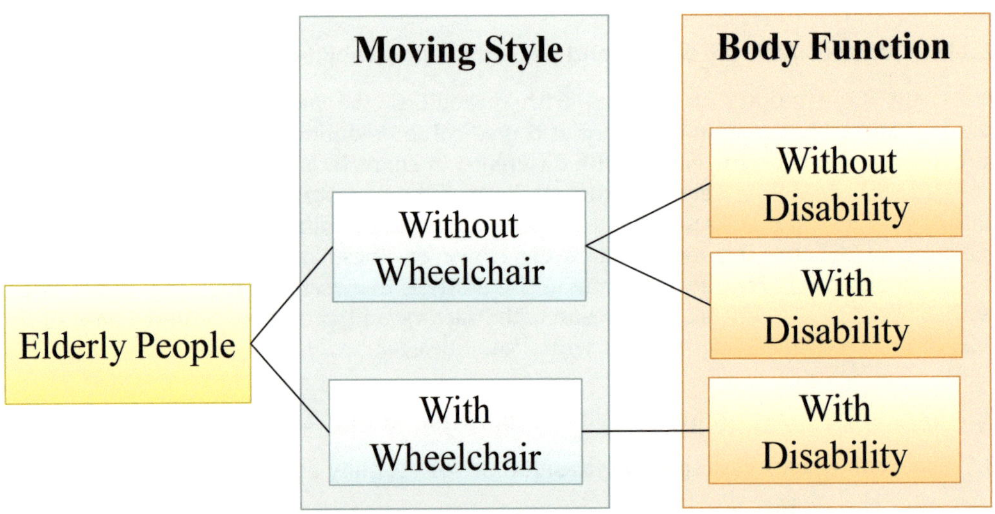

Fig. 2. Classification of survey participants (M. Nihei, et al, 2007).

Mobility Aid	Device Function	Body Function
 Walking stick	Walking-aids with one leg and a handle without forearm support.	Person for whom support when walking is necessary. The person grasps the stick to support their weight.
 Rollator	Wheeled frames with handgrips and legs that provide support whilst walking.	Person who needs greater support and stability than that provided by a walking stick.
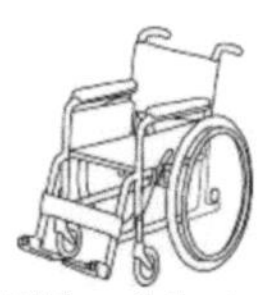 Wheelchair	Wheelchairs designed to be propelled by the user, by pushing with both hands on the rear wheels or on the hand rims of the rear wheels.	Person who can walk for short periods.
 Electric motor-driven wheelchairs with manual steering	Wheelchairs powered by electricity that are steered by direct mechanical linkage without power assistance.	Person who cannot walk for long periods.

Table 1. Typical personal mobility aids in daily living in Japan (Association of Technical Aids, 2002).

3.3 Psychological conceptual model for the introduction of mobility aids

The psychological situations of elderly people were clarified by visualization based on the investigation results, and the relationships among these factors were derived through a psychological process based on antinomy. There are two main tradeoffs with respect to ATs, namely that between body function and mobility, and that between body appearance and effortless mobility. For example, if their physical functions decline, persons with disabilities chose to use a mobility aid. Such devices enable them to extend their area of movement. However, a further decline in physical function may occur. This study clarified the psychological states of elderly persons when trying to resolve these dilemmas.

3.4 Assistance with mobility activities

Based on our research findings, we built a conceptual model that AT developers can easily understand. In the model, we paid attention to users and ATs, mobility in daily life, focusing on daily mobility among elderly persons who used/had not used/had experienced using a wheelchair at home. We emphasized two viewpoints: mobility by wheelchair and

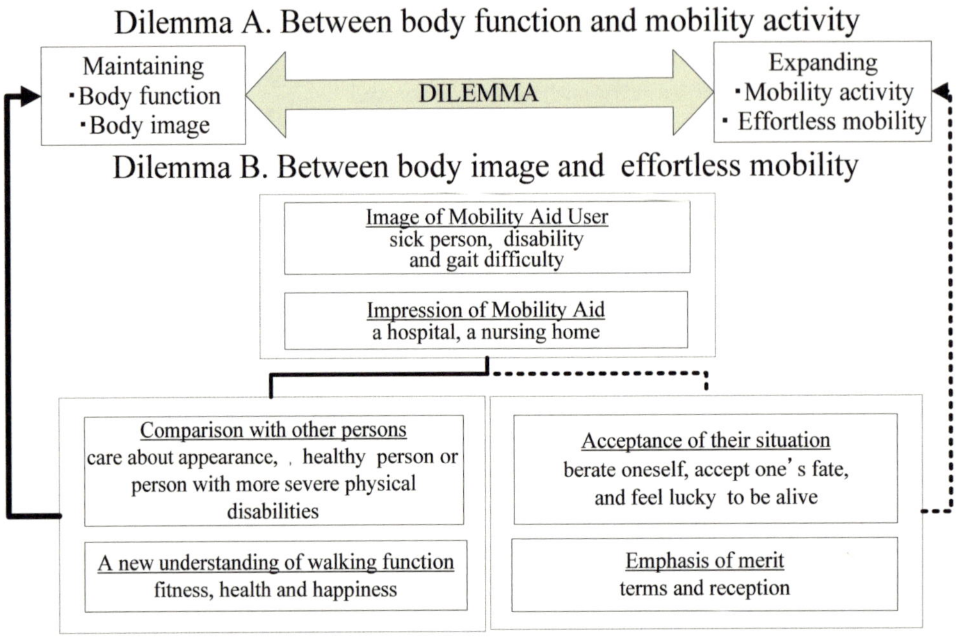

Fig. 3. A dilemma-resolving process model for elderly persons (M. Nihei, 2008).

mobility by walking as viewed by wheelchair users and nonusers. We concluded that: wheelchair users are satisfied with the status quo, but initially had psychological conflicts. Former wheelchair users who are now walking fear physical deterioration, while at the same time recognizing the usefulness of wheelchairs. Acceptance by the elderly of wheelchairs depended on their impressions of and resistance to wheelchairs. These impressions were explained by a disability-acceptance phase model. Wheelchair nonusers clearly harbored some resistance to wheelchairs. We found that the psychological conflicts identified in our survey involved: (a) anxiety about deterioration in physical functions due to AT use, (b) expansion of mobility thanks to AT use, and (c) wheelchair use augmenting an image of physical deterioration and disability. We conceptually diagrammed these psychological states, illustrating the mechanism of psychological resistance and conflicts in a trade-off relationship.

3.5 Assistance of mobility and ICF

What is the best way to resolve these dilemmas? The goal of rehabilitation is the medical viewpoint (medical rehabilitation), such as a disability has to be cured and depression of body function was the only cause for disability for a long period. Therefore, they spend a long rehabilitation time to cure the impaired function. Since 2001, ICF (International Classification of Function), which was established by WHO (World Health Organization), consider the goal of rehabilitation to be body structure and function, activity and participation (WHO, 2001). This representation was an important turning point in stating that rehabilitation is not only a medical issue, but involves but also daily activities and social life.

Here, we discuss the results obtained in Section 3 and the modality of mobility. The idea of selecting physical therapy and non-use of AT to avoid physical depression caused by the use of a wheelchair is an idea based on medical rehabilitation. On the other hand, the idea of selective use of a (powered) wheelchair is an idea based on social rehabilitation. These ideas are poles apart, and it was considered that these ideas were incompatibility. Is that true? In this paper, we propose a new solution to resolve satisfy both medical and social rehabilitation.

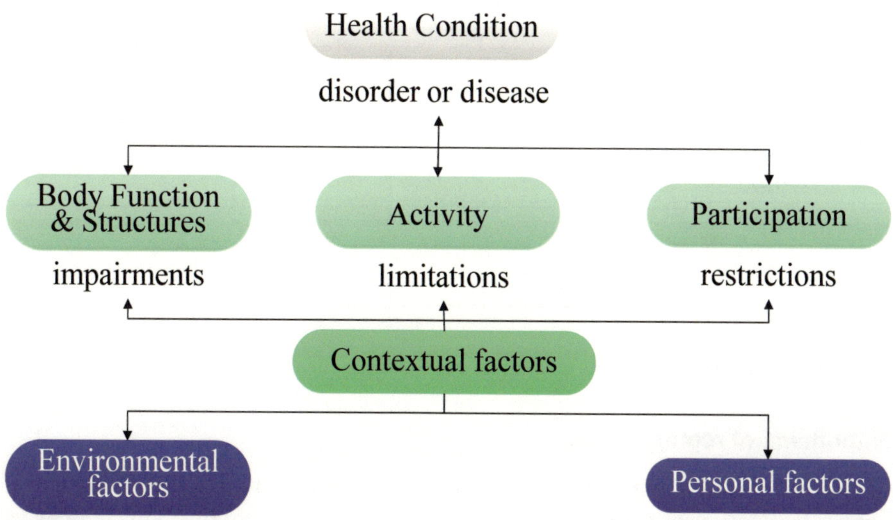

Fig. 4. The International Classification of Functioning, Disability, and Health (WHO, 2001).

4. Concept of a new mobility

4.1 Developmental requirements (M. Nihei et al., 2006-2008)

Our novel concept was created based on the model described above. The cause of psychological conflict of elderly people who have to use a wheelchair: "decline of physical functions," "reduction in mobility activities," "resistance to being a person with a disability," were clarified by a qualitative research approach. Proposing development requirements and introducing their life is needed in order to reduce the psychological conflict and to devise a solution for the situation of elderly people who have to use a wheelchair. To this end, three developmental requirements were proposed and defined.

- Maintain body function: maintain muscles and joints motion for walking.
- Expand mobility activity: expand mobility activity related to daily activities and participation
- Maintain natural walking appearance: maintain a natural walking style and gait

This concept includes the following three functions: enabling natural walking, similar to walking on flat ground; amplifying movement speed and enabling comfortable movement; and maintaining the appearance of being able to walk. Table 2 shows the estimation of the psychological conflict against these mechanical functions of existing products.

The targeted persons have the following characteristics: the ability to walk under their own power, but not quickly; the ability to maintain or expand their mobility area; and the appearance of being able to walk.

Required function	New concept	Walking stick, Rollators	Wheelchair	Powered wheelchair	Trainer
Maintaining body walking function	○	○	×	×	○
Expanding mobility activities	○	△	○	○	×
Maintain natural walking appearance	○	○	×	×	×

Table 2. Comparison between new concept and existing products.

4.2 Components of requirements

To decide upon a more concrete mechanical requirement than that listed above, we initially categorized the relationship between the human bodily phase and mechanical phase shown in Fig. 5. Human functions consist of main three factors: the sensory system, the central nervous system, and physical function. These three factors are integrated in body movement and bodily appearance is dictated by movement. Likewise, the mechanical system consists of main three factors: the operation system, the control system, and mechanical function. These three factors are integrated in mechanical design. The interfaces that connect human and mechanical function are the human sensor receptor to recognize the mechanical function (behavior) and the operation system to use the physical function. Therefore, to meet the developmental requirement, amplifying the results of the system's movement for the operation of one's own body movement is one solution.

Second, using a combinational method (H. Yoshikawa, 1985) with keywords and partial functions, solutions were extracted by using the matrix shown in Figure 6. The developmental requirements are shown in the following equation.

$$T = T^1 \cap T^2 \cap T^3 \tag{1}$$

Here, T is developmental requirement, T^1 is maintain body function, T^2 is expanding mobility activity, T^3 is maintaining a natural walking appearance. First, we extracted a keyword for each factor, e.g. walking, treadmill, training machine, stepping machine in the case of T^1, and the flow of the developed figure. Here, S_1 is the expected duration of product development, S_2 is the feasibility of technical issue.is shown in Figure 6. Based on one example of these solutions, we propose a new mobility aid.

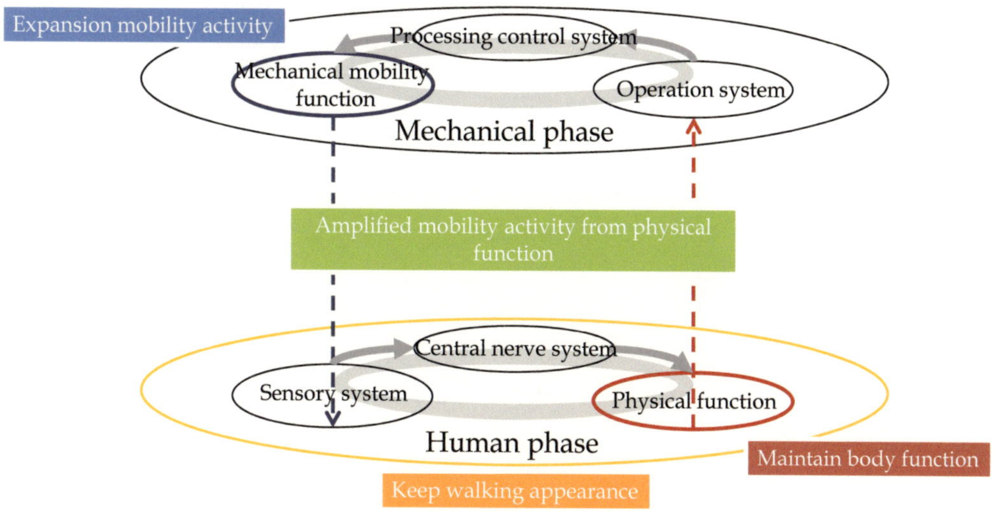

Fig. 5. Development requirements between bodily and mechanical phase.

4.3 New mobility, tread-walk (M. Nihei et al., 2006-)

We have designed and developed a new mobility-aid for the elderly that improves their walking speed, as shown in Figure 7. This device is called Tread-Walk1 (TW-1). It is mobile treadmill-like system that is designed to meet the following three requirements: the maintenance of bodily function; the extension of mobility; and the maintenance of the appearance of a natural gait. Targeted users have the following characteristics: the ability to walk under their own power, but not quickly; the motivation to maintain or expand the geographical area in which they are mobile; and the wish to appear as though they are walking naturally. The control of the system's belt velocity is based on the natural walking velocity of the user. The user's intended walking speed is detected by the kicking and braking forces applied by the user's feet as he or she walks on the treadmill belt, which in turn is connected to a DC motor. These forces are counteracted by the propulsion force of the belt. To control this system more accurately, signals for rotational movement should also be derived directly from the walking motion. In this section, we focus on human rotational movement, and we describe a novel system that uses two belts to derive separate signals from each foot during walking.

Figure 8 shows a moving walkway. The developed system, the TW-1, gives the same feeling as walking on a moving walkway; just imagine the marvelous feeling of acceleration.

4.4 Prototype construction

1. Basic concept

The mobility aid shown in Figure 9 is a four-wheeled vehicle that permits the user to walk naturally, while a servomotor amplifies the natural walking speed of the user. Its main components are a treadmill and two front driving wheels. Sensors in the treadmill detect the acceleration/deceleration forces applied to the surface while walking. The rotational speed of the driving wheel motors is controlled by signals from these sensors. Computer software

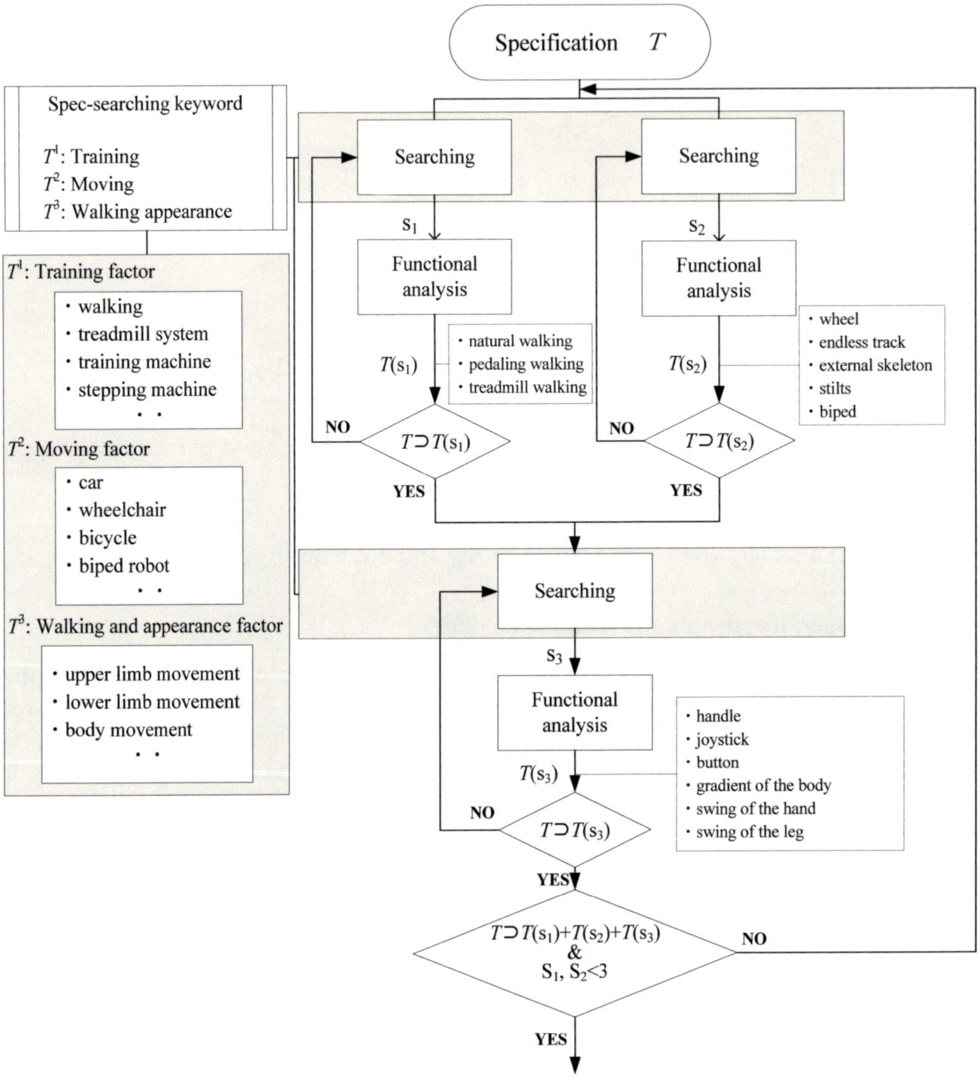

Fig. 6. Development requirements between bodily and mechanical phases.

Fig. 7. Artists impression of Tread-Walk1.

Fig. 8. Image showing the pleasant walking feeling of Tread-Walk1.

monitors the walking pattern, as measured by the treadmill motor load, and the steering control detects the angle of the steering wheel and controls the speed ratio of the right and left wheels.

Figure 10 shows the system flow of the functional overview of the treadmill and the driving parts. The treadmill motor actuates the treadmill belt and acts as a sensor device at the same time. The mobility aid drive operates as follows:

i. The kicking force of the user rotates the treadmill belt
ii. The rotation force is directed to the shaft, and the load current is detected as kicking and braking forces while walking
iii. The increase or decrease in the rotation speed is decided in the same manner as the kicking or braking force, based on the current load signal, which is derived by a computer program
iv. The mobility aid operates synchronously with the treadmill belt, but the velocity is increased by driving motors.

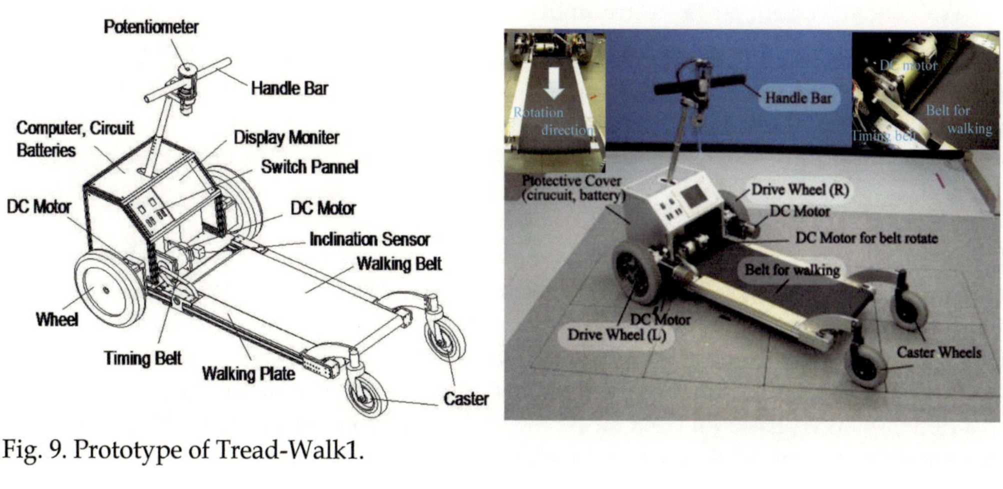

Fig. 9. Prototype of Tread-Walk1.

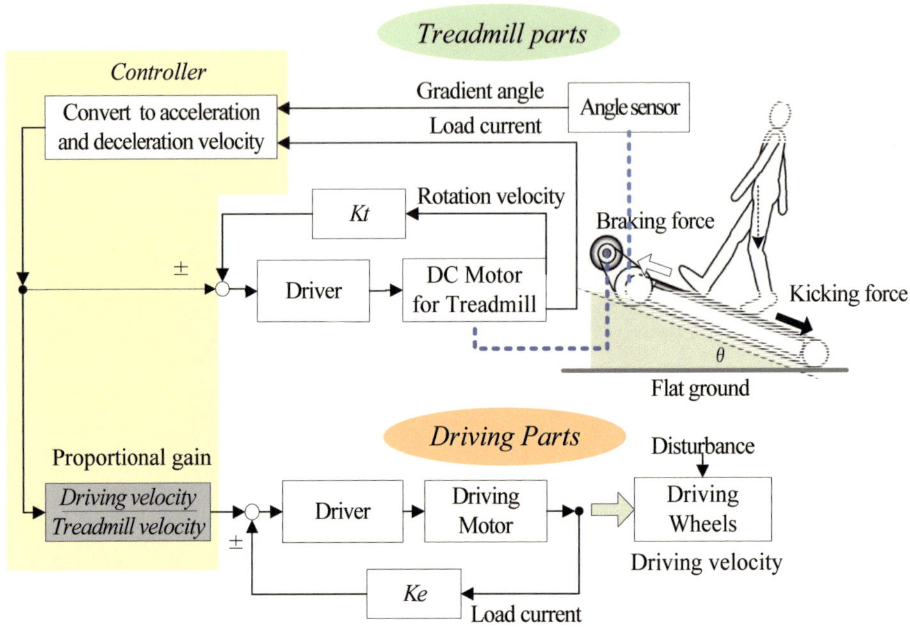

Fig. 10. The system flow of the functional overview of Tread-Walk1.

2. Design for the elderly

The vehicle's external dimensions are 1520 (L)x760 (W)x1120 (H) [mm] and the maximum velocity is 6.0 [km/h], and are reference from JIS (Japanese Industrial Standardization). The size of each component such as platform height, withstanding load, continuous operation time, handle length and height, grip size, rotational resistance of handle, and size of treadmill were determined by consulting a database of human body dimensions and data on the characteristics of the body functions of the elderly (NEDO, 1993) (shown in Figure 11). The specifications were also determined by taking into consideration envisioned risks such as the likelihood of elderly users falling off the vehicle.

3. Walking movement on TW-1 treadmill

Control of the treadmill belt velocity was based on the natural walking velocity of the user. The user's intended walking speed is detected by the kicking and braking forces that are applied by the user's feet as he or she walks along the treadmill belt, which is connected to a DC motor. These forces are counteracted by the propulsion force of the belt as load trque.

4. Vehicle movement and acceleration

The technological opportunity, the walking assist-rate, assist force, slope assistforce and acceleration were extracted. The vehicle control mthodology was designed to be safe, taking into consideration the inertial force resulting from the acceleration of the vehicle. Figure 11 shows the gait pattern during walking on Tread-Walk1 and movement, recorded by video capture every 0.2 [s]. These photographs shows that the gait is a natural pattern including (a) a double-stance phase, (b) push off by the left foot, (c) foot off by the left foot, (d) stance phase of right foot and (e) foot contact of left foot. (T. Ando et al., 2009)

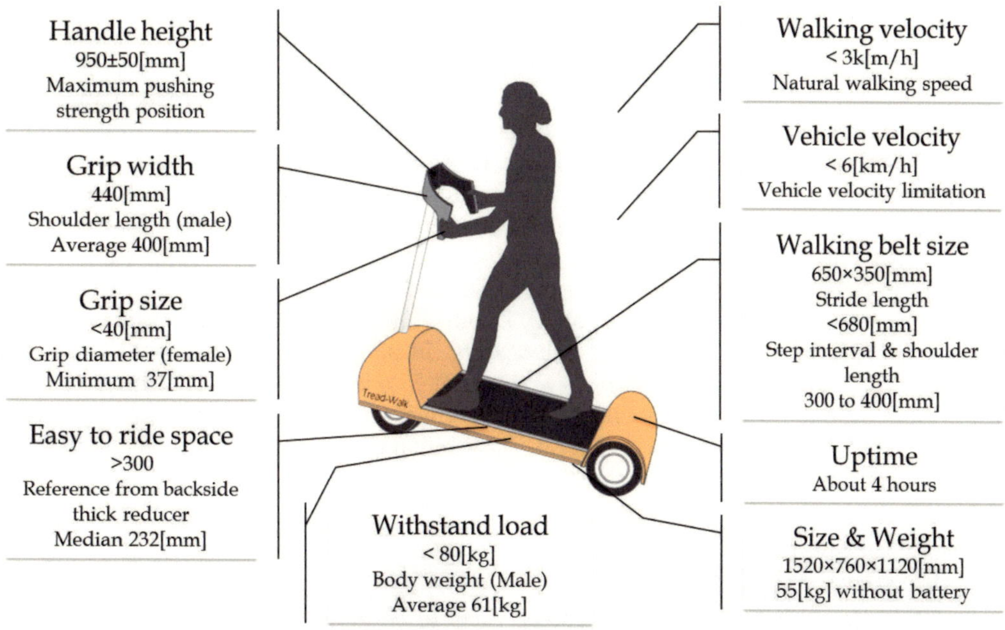

Fig. 11. Specifications and design for the elderly.

5. Vehicle steering

We propose a method for steering control that enables the user to turn the device safely and in a stable manner. The system is designed to allow the user to maintain his or her balance in a standing position during a turning operation. To achieve this, the rotational velocity at which a comfortable balance can be maintained in a standing position when the user is subjected to a centrifugal force was established (T. Ando et al., 2008).

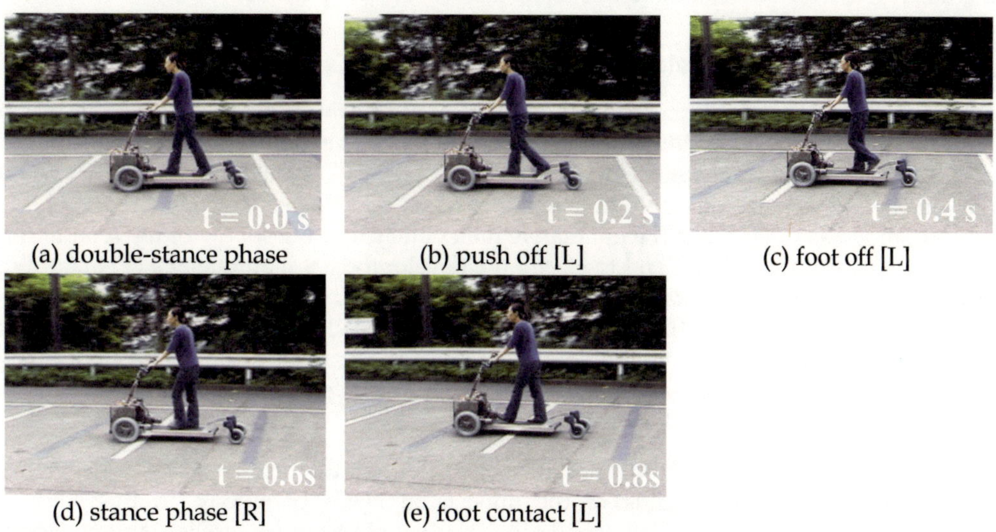

Fig. 12. Gait pattern during walking and driving Tread-Walk1.

4.5 Evaluation of prototype and impression of new mobility aid

Test drive evaluations in an open space suggest that this device is safe and effective for use by middle-aged and older adults. Walking movement on TW-1 was analyzed kinematically by comparing it with the corresponding walking movement on flat ground. To determine whether the two dilemmas between body function and mobility activity, and body appearance and effortless mobility of the mobility aid had been addressed, detailed interviews were conducted to determine the impressions of users regarding the proposed mobility aid. We interviewed 38 healthy adult participants aged over 60. We performed test drives in an open space, and each driver demonstrated their test drive to other participants. The question "How do you feel about using Tread-Walk1?" was asked, and responses were made according to a five-grade evaluation scale. In addition, free comments were also recorded. The results of the interview revealed that 79% of the participants had a positive impression of the proposed mobility aid. Comments such as "interesting," "exciting," "novelty item," "good for health," "exercise," "rehabilitation, " and "increased willingness to go out," were provided by the participants.

(a) Participant A (b) Participant B

Fig. 13. Test driving and evaluations, subject (healthy elderly persons).

4.6 Proposal for a new development methodology for ATs

Figure 14 shows the development methodology used in this study. The first point was to introduce participant observation (the Cogitation stage) and a qualitative research approach at an early stage (the Analysis stage). This methodology can reveal latent needs that do not rely on the individual experience and thinking of users, key persons, and therapists. The second point was to visualize the qualitative data obtained in the survey of the relationship between ATs and the elderly, and to clarify the developmental requirements (Visualization of the tradeoff relationship stage). The result of the qualitative research was translated to a conceptual model for visualization and the developmental requirements were extracted from the main components of the conceptual model. To decide the shape of the new mobility aid, a components matrix and human-machine system were introduced because there are several solutions for the design.

A new mobility aid, Tread-Walk1, was developed through the proposed methodology that was based on an understanding of the undeclared needs of elderly. In this study, a prototype of Tread-Walk was developed with the goal of resolving conflicts among elderly people regarding the use of AT. According to a survey based on the qualitative research approach, this conflict was described schematically as a dilemma between a willingness to maintain-improve body function and an extension of the mobility area. To resolve this conflict, we proposed the concept of a new mobility aid. In the result of this study, the impression of the mobility aid was reported by elderly people to be positive. Psychological issues, such as a resistance to the wheelchair, which clearly emerged in this study, is a real and substantive problem. In addition, the development of useful AT will be improved by taking into account a composite observational study of the bodily, living and psychological situation of elderly people. While the model we developed does not apply to all elderly people, it can be treated as a representative case of a realistic situation.

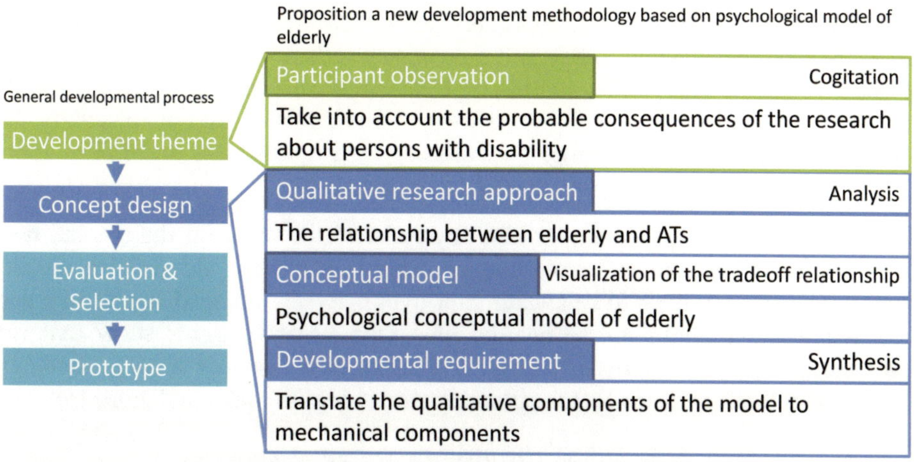

Fig. 14. Development methodology using a qualitative research approach.

5. ATs and an aging society

5.1 Additional issues

As addition issues, we surveyed the elderly and people with disabilities living in nursing homes who routinely use wheelchair as to AT satisfaction and the psychological effects. As a result, the elderly were almost all pleased that more people with disabilities were using ATs and their services. However, lower overall psychological effects were found to be less aggressive, especially "Adaptability" score.

While the elderly are satisfied with the AT that is available; it was shown that there is a potential psychological effect is difficult to obtain due to the AT. Considering these results from a survey, we conducted in-depth interviews at the same time "(Even if I get more advance AT), there is no place to go," and can dilute the purpose of such movement, "I don't feel like a new things now" and lack of purpose in life and found that the presence and movement activities are meaningless. These result are what leads to non-use of AT, to provide a useful instrument for the elderly is truly what we do, a comprehensive approach is necessary for life beyond the framework of the high life and considered.

5.2 The role of technology for an aging society

Here, we discuss how the positioning of the ATs in this study contrast with the background of an aging society and ideology. There are mainly two different ideologies between the elderly and ATs. One ideology focuses on improving QOL by promoting physical health leading to comfortable life—coincidentally a mainstream social policy. Maintaining physical health reduces medical expenses and lightens the care burden, together with public financial and physical burdens. Extending this ideology too far, however, may lead to excessive "embodiment (L. Cheryl, 2003) " centering on an ideal body image rather than the realities of the aging body itself. This ideology in engineering is the mainstream, as it has been predicted to increase in the number of consumer AT to the increasing number of elderly people, and commercialization has developed various devices entered many enterprises have been promoted.

The other ideology involves an ideal called "empowerment (C.M. Morell 2003)," and is based on the idea that the will to live and a positive mindset will improve QOL, regardless of physical function. This values neither age nor physical function, focusing instead on spontaneous self-motivation, making it easier for individuals to "accept" disability, illness, or death. In the field of engineering, they are powered-wheelchairs and information and communication technologies.

These two ideologies, one focusing on the physical and the other on the mental, both are likely to improve QOL, but their components differ. Most of the elderly thus are finding their own optimum solutions to balancing their lives between "an aged but healthy society" and " a wonderful thought not so healthy life." Our research results show that we can improve the QOL of the elderly by using AT. Instead of suggesting ways of ideology, we propose developing ATs based on the potential needs of elderly people (Figure 15).

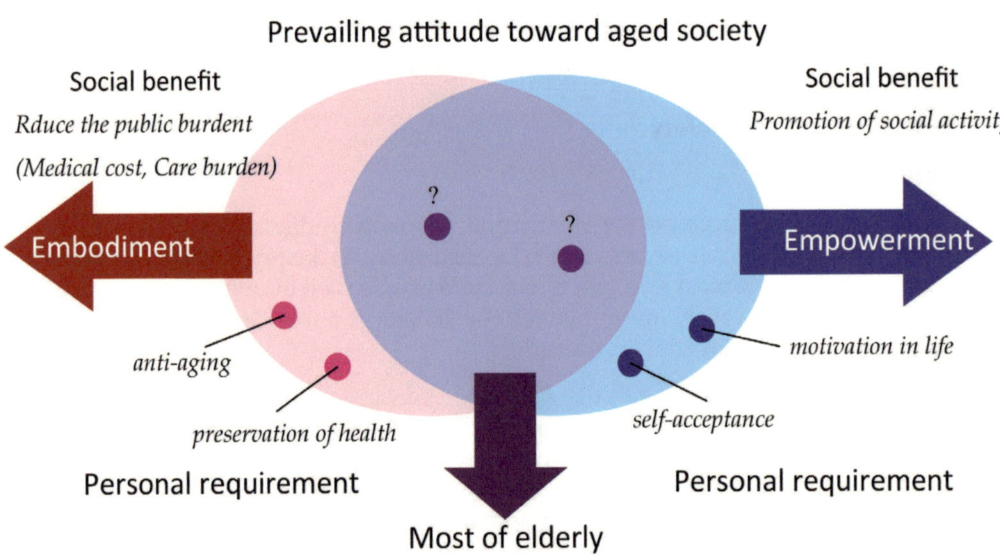

Fig. 15. The elderly and social thought.

6. Conclusion

One of the outcomes of this research is that qualitative research of the psychological trends of individuals who use assistive technologies clearly exists in regard to the relationship between users and assistive technologies, which was clarified by an analysis of this concept, and these findings are expected to be applied to rehabilitation engineering and clinical fields in the future. In addition, this research can greatly contribute to future progress in rehabilitation engineering by providing a concrete example of device development, by showing the necessity of considering user psychology during development, and by proposing a new methodology for use in development.

7. Acknowledgements

We thank T. Inoue, Research Institute of National Rehabilitation Center for the Person with Disabilities for cooperation in this study and participants for his invaluable contributions to insights in this work. And C. Yamaki, Y. Kaneshige, T. Ando, M. Y. Nakashima, N. Yamauchi and M. Mieko for their participants for their invaluable contributions to insights in this work. This work was supported in part by KAKENHI 207004600.

8. References

Ando, T., Nihei, M., Kaneshige, Y., Inoue, T., Fujie, M.G. (2008). A Steering System of a New Mobility-Aid Vehicle with walking: Tread-Walk, *The second IEEE RAS/EMBS International Conference on Biomedical Robotics and Biomechatronics.*

Ando, T., Nihei, M., Ohki, E., Kobayashi Y., Fujie, M. G. (2009). Kinematic Walking Analysis on a New Vehicle Tread-Walk with Active Velocity Control of Treadmill Belt, Conf *Proc. IEEE Eng Med Biol Soc*, 5977-598.

B. Louise, J. Kim, B. Weiner, (2002). The Shpaing of Individual Meanings Assigned to Assistive Technology: a review of Personal Factors, *Disability and Rehabilitation 24*, No.1/2/3, 5-20.

Body Function Database of elderly persons, http://www.hql.jp/project/funcdb1993/(ref. 2011-7)

C.M. Morell, Empowerment and long-living women: retuen to the rejected body, Journal of Aging Studies, Vol. 17, No.1, pp.69-85, 2003.

The Cabinet Office, an aging society white paper 2011, 2011 edition, http://www8.cao.go.jp/kourei/whitepaper/w-2011/gaiyou/23pdf_indexg.html (ref. 2011-7)

D. Lupton, et al., Technology, Selfhood and Physical Disability, Social Science & Medicine, 50, pp.1851-1862, 2000.

DJ. Baker, D. Reid, C. Cott, (2004). Acceptance and Meaning of Wheelchair use in Senior Stroke Survives, *American Journal of Occupational Therapy*, vol. 58-2, 221-230.

H. Rosalind et al., (1998), Wheelchair Users and Postural Seating - A Clinical Approach-, *Curchill Living Stone.*

H. Yoshikawa, (1985). General Design Tehory, Mechanical Science Vol37-1, 108-116.

J. Jutai, et al. (1999). Quality of life impact of assistive technology. *Rehabilitation Engineering, RESJA (in Japanese)*, 14, 2-7.

Japanese Industrial Standarization, Motored wheel, JIS T9203, 2003.

Y. Kaneshige, M. Nihei, M.G.Fujie, (2006). Development of New Mobility Assistive Robot for Elderly People with Body Functional Control -Estimation walking speed from floor reaction and treadmill-. *Proceedings of the IEEE RAS-EMBS*, 79.

K. Masahiro, (2005). For the Creation of New Market of the Aged Society with a Fewer Number of Children, *Journal of the Japan Society of Mechanical Engineers*, Vol 108, No.1038, 37-42.

L. Cheryl Laz, Aged embodied, Journal of Aging Studies, Vol 17, pp.509-519, 2003.

R. Morales, et al., (2006). Coordinated Motion of a New Staircase Climbing Wheelchair with Increased Passenger Comfort, *Proc. of the 2006 IEEE ICRA*, 3995-4001.

S. Hashimoto, (2007). ISO/IEC Guide 71 and Physiological Characteristics of Japanese, *Rehabilitation Engineering*, Vol.22, No.3, 144-150.

T. Inoue et al., (2002). Development of Head Controlled Powered Wheelchair Based on Components of QOL, *J of the Japanese Society for Wellbeing Science and Assistive technology*, Vol.1-1, 42-49.

M. Roelands, et al., (2002), A Social-Cognitive Model to Predict the Use of Assistive devices for Mobility and Self-Care in Elderly People, *The Gerontorologist*, Vol 42-1, 39-50.

The Association for Technical Aids (2003), Technical aids for persons with disabilities - Classification and terminology-, *ATA*, 53-58.

Murray et al. (2003), Modeling of a stair-climbing wheelchair mechanism with high single step capability, *IEEE Trans. On neural system and rehabilitation engineering*, Vol.11, No.3, 323-332.

M. Watanabe et al., (2006). Outlie of the Strategic Technology Road Map of METI and Expecation for the Society of Life Support Technology and the Japanese Society for Wellbeing Science and Assistive Technology, *J. Of the Japanese Society for Wellbeing Science and Assistive Technology*, Vol.6, No.1, 3-12.

NEDO, Physical function database of elderly people, walking width in free walking, (Available 2007.2) : http://www.hql.jp/project/funcdb1993/.

M. Nihei, Y. Kaneshige, M.G. Fujie, T. Inoue, (2006). A Hyrid Manual/Motorized Mobility Device for Asssisted Walking, *Biomechanism 18* (in Japanese), 101-111.

M. Nihei., Y. Kaneshige, M.G. Fujie, T. Inoue, (2006). Development of a New Mobility System Tread-Walk -Design of a Control Algorithm for Slope Movement-, *Proceedings of the 2006 IEEE International Conference on Robotics and Biomimetics*, 1006-1011.

M. Nihei, Y. Kaneshige, T. Inoue, M.G. Fujie, (2007). Proposition of a New Mobility Aid for Older Persons –Reducing psychological conflict associated with the use of Assistive Technology, *Assistive Technology Research Series*, AAATE 07, Vol.20, 80-84.

M. Nihei, T. Inoue, M. Mochizuki, C. Yamaki, T. Kusunaga, M. G. Fujie, (2007). Proposition of Development Concept of Mobility Aids Based on Psychological Model of Older Persons, Transactions of the Japan Society of Mechanical Engineers, Series C, 73-725, 266-273.

M. Nihei, T. Inoue, M. G. Fujie, (2008). Psychological Influence of Wheelchairs on the Elderly Persons from Qualitative Research of Daily Living", *J. of Robotics and Mechatronics Vol.20 No.4*, 641-649.

M. Nihei, T. Ando, T., Y. Kaneshige, M.G. Fujie, T. Inoue, (2008). A New Mobility-Aid Vehicle with a Unique Turning System. *Proceedings of the 2008 IEEE/RSJ IROS*, 293-300.

M. Nihei, T. Ando, Y. Kaneshige, T. Inoue and Masakatsu G. Fujie (2010). Development of a New Vehicle Based on Human Walking Movement with a Turning System, *Robotics 2010 Current and Future Challenges, Houssem Abdellatif (Ed.)*, ISBN: 978-953-7619-78-7, INTECH.

WHO, (2003). International Classificatioon of Functioning, Disability and Health, *Chuohoki*.

Y. Shinomiya (2001), Development of Horseback riding Therapeutic Equipment and its Verification on the Effecton the Muscle Strength Training, *TVRSJ*, Vol.6, No.3, 197-202.

Zeng, Q., Teo, C. L., Rebsamen, B., Burdet, E. (2006). Design of a Collaborative Wheelchair with Path Guidance Assistance. *Proceedings of the 2006 IEEE ICRA*, 877-882.

8

New Strategies of Mobility and Interaction for People with Cerebral Palsy

R. Raya, E. Rocon, R. Ceres, L. Calderón and J. L. Pons
Bioengineering Group - CSIC
Spain

1. Introduction

Cerebral palsy (CP) is one of the most severe disabilities in childhood and makes heavy demands on health, educational, and social services as well as on families and children themselves. The most frequently cited definition of CP is a *disorder of posture and movement due to a defect or lesion in the immature brain*, Bax (1964). The prevalence of CP is internationally 1.5-2.8 cases per 1000 births. Only in the United States 500,000 infants are affected by CP, Winter et al. (2002). In Europe these figures are even higher, Johnson (2002).

The "Surveillance of cerebral palsy in Europe: a collaboration of cerebral palsy surveys and registers" presented a consensus that was reached on definition of CP, classification and description, Cans (2000). This classification divides CP into spastic, ataxic and dyskinetic.

It is demonstrated that it is during early stages of development that fundamental abilities and skills are developed, Gesell (1966); Hummel & Cohen (2005). Thus, it is crucial to give CP infants an opportunity to interact with the environment for an integral development. It is recognized that assistive technology can improve the functional capabilities limited by CP, LoPresti et al. (2004). Approximately 50% of children affected by CP need technical aids for assisting their mobility (braces, walkers, or wheelchairs).

The motivation of this work arises from the limitations caused by CP in the fundamental areas of human being: mobility, communication, manipulation, orientation and cognition. The aim of this work is to design assistive devices to promote the fundamental skills of children with CP reducing motor and cognitive limitations.

One the one hand, the capacity of exploring and controlling the environment is essential for any human being. Independent mobility plays a crucial role in this exploration, leading to the child's physical, cognitive and social development, Azevedo (2006). According to the state of art of mobility devices for people with CP, most devices are more focused on mobility than the integral development.

One the other hand, the posture and motor disoders associated to CP limit frequently the access to general assistive devices. The human machine interaction results a critical factor. According to the state of art of person-computer interfaces for people with CP, there are a wide diversity of solutions. However, authors assert that the usability decreases dramatically when users have a severe motor disability.

From such considerations, the following needs are identified: 1) To emphasize the learning of the physical and cognitive skills through mobility experiencies, 2) to characterize the control limitations caused by posture and motor disorders, 3) to facilitate the interaction between the child and the assistive device by filtering those control limitations.

In this context, the main contributions of this work are:

- Design and evaluation of a robotic vehicle, called PALMIBER, for cognitive and physical rehabilitation.
- Software tool for objective evaluation of the driving task and interfaces.
- Design and evaluation of a novel person-machine motion-based interface, called ENLAZA, to increase the accessibility of the vehicle.
- Characterization of the posture and motor disorders of people with CP by using the ENLAZA interface.
- Design and validation of a filtering technique to reduce the effect of the involuntary motion on the control of the interface.
- Functional validation of the ENLAZA interface and filtering technique as pointing device for the computer.
- Functional validation of the ENLAZA interface as input device for driving the vehicle.

A review of asssitive devices for mobility focused on people with CP is presented in section 2. Section 3 presents the design and evaluation of the vehicle PALMIBER, a pre-industrial prototype for the integral rehabilitation. A review of person-computer interfaces for people with CP is presented in section 5. The ENLAZA interface is presented in section 5.1. Section 5.2 describes the characterization of motor disoders of people using the ENLAZA interface. Section 6 presents a novel algorithm to reduce the effect of the involuntary movements on the control of the interface. Finally, section 7 presents the functional validation of the ENLAZA interface as input device for the computer. Section 8 presents the functional evaluation of the ENLAZA interface as input device for driving the PALMIBER vehicle.

People with CP and therapists from ASPACE Cantabria (Spain) have participated in all phases from the conceptualization to the validation to guarantee the usability of the devices. The inclusion criterion has included people with severe motor disorders which have been described. It is expected that results are representative for people with similar profile.

2. Assistive devices for mobility focused on CP

There are basically two groups of assistive devices to help people with mobility problems: the alternative devices and the empowering (or augmentative) devices. These solutions are selected based on the degree of disability of the user. In the case of total incapacity of mobility (including both bipedestation and locomotion), alternative solutions are used. These devices are usually wheelchairs or solutions based on autonomous especial vehicles. Because of that, the wheelchairs are the focus of research of many groups all over the world.

Advanced wheelchair-based devices constituting a doctrine body known as Autonomous Robotic Wheelchairs (ARW), which can be considered a specialized version of the well-known autonomous mobile robot (AMR). In fact, a number of topics are common to both technologies, particularly obstacle detection and avoidance, localization, path planning, and wall following. Nevertheless, an important distinction must be made because an ARW is specifically designed to transport people. As a result, some issues become very important, such as safety, stability, smooth driving, and higher maneuverability, generally due to nonstructured environments. Many references can be found in the literature about research projects leading to experimental prototypes, such as OMNI, Hoyer et al. (1997), SIAMO, Mazo et al. (2002), VAHM, Bourhis & Agostini (1998), TetraNauta, Civit (1998), BREMEN, Rofer & Lankenau (1998), and SENARIO, Katevas et al. (1997).

Different technical aids focused on CP can be found such as canes, walkers, orthosis. Canes are the most traditional tool to help people walk. Walkers are a device to maintain balance or stability while walking. There are many types of walkers, considering their constitutive materials, accessories, sizes and structural configurations. An orthosis is an external orthopedic appliance that prevents or assists the movement of the spine or limbs. There are some interesting solutions which combine both solutions.

The NF Walker by EO Funktion® has been developed to give children the possibility for movement, both standing and walking. It promotes body awareness and improves respiratory and cardiovascular functions through regular activity, as well as reduces the risk of chronic diseases such as high blood pressure, osteoporosis and gastrointestinal disorders. The SMART Walker® Orthosis has been designed and manufactured by Advanced Orthotic Designs. It encourages a child with CP to learn to stand and ambulate with hands-free support.

There are assistive devices for motor rehabilitation which combine orthoses, user's weight support and a treadmill. The Lokomat®, designed by Hocoma, is a driven gait orthosis that automates locomotion therapy on a treadmill and improves the efficiency of treadmill training. Preliminary studies have demonstrated that Lokomat® has a positive impact on motor rehabilitation of children with CP, Borggraefe et al. (2009).

Most of the implementations described above are conceived as technical aids to help increase the mobility of the disabled. A few of the systems described in the literature address the particular problems of children affected by neuromotor disorders accompanied by mental retardation. One outstanding case worthy of mention is the Communication Aids for Language and Learning (CALL) center that developed a smart wheelchair specifically for children with mobility impairment. In this case, a standard wheelchair was instrumented and adapted in terms of the user's interfaces.

The *Gobot* is a special vehicle that enables mobility for children with CP created by the Hospital of Lucile Packard Stanford (US). It moves easily from a horizontal to a vertical position. Tray, hip, lateral supports and full foam knee supports are also simple to adjust. The *Magellan Pro Robot* is a commercial robot made by the iRobot corporation. Researchers from the Delaware University (US) used it for rehabilitation of children with CP, Galloway et al. (2008); Lynch et al. (2009). The robot was equipped with an on-board computer and odometry. Their results demonstrated that young infants will independently move themselves via a mobile robot. Their data do provide indirect evidence that infants were not simply focused on moving the joystick but were associating joystick activation with their motion.

Although these devices are focused specifically for children with CP, they are based on standard or commercial devices. The following work addresses the development of an assistive platform designed specifically for alternative mobility of children affected by CP. Both mechanical structure and driving modes have been designed according to the user's needs. Moreover, the vehicle PALMIBER fulfills the technical requirements related to safety,

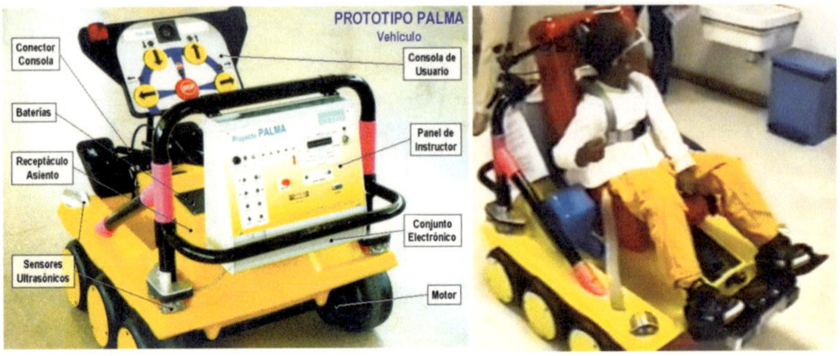

Fig. 1. Prototype PALMA project

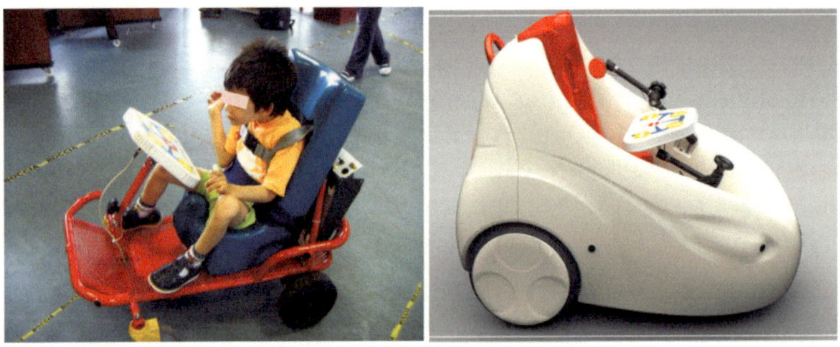

Fig. 2. Prototypes of PALMIBER project

stability and robustness. This rehabilitation tool is conceived as an attractive tool for integral development of children with severe neurological problems.

3. The PALMIBER vehicle

The origin of the PALMIBER vehicle (Fig. 2) was the PALMA project, Fig. 1. The PALMA project was developed with the main aim of proposing an attractive tool for integral development of children with severe neurological problems, Azevedo (2006); Ceres et al. (2005). PALMIBER vehicle aims to create a commercial product based on the concept validated by PALMA project.

3.1 Driving modes

Regarding system driving autonomy, the basic requirement is the adaptability to user dexterity and to user progress and mental development. As a consequence, our approach is basically different to the one presented for the previous ARWs. We are interested in improving children's motor control, driving skills, and their ability to eventually reach a desired destination. The educators in the PALMA team identified six steps in the cognitive development of the children. They are summarized in table 1.

3.2 Electronic architecture and peripheral devices

- A dsPIC controls the different modules of the PALMIBER: the user and educator interfaces, odometry module, obstacle detection system and power and motor management.

Driving mode	Description
Autonomous (Level 0)	The vehicle detects and avoids obstacles without user intervention. In this navigation mode, the vehilce navigates while avoiding obstacles without any predefined target.
Cause-effect relation (Level 1)	The child press any key and the vehicle start to move. The vehicle stop when it finds a obstacle.
Training of the direction (Level 2)	The vehicle stops after detecting an obstacle (a crashing noise and alarm light warn of the eventual crash). The child must press the correct button proposed by the vehicle otherwise the vehicle does not move. If no response is obtained from the child, he is invited to do so by a verbal message.
Decision of the direction (Level 3)	The user decides and presses the driving buttons, but the vehicle automatically stops if any obstacle is detected. Verbal commands are generated. At this level, the children start deciding on a navigation target.
Fully user guided (Level 4)	The vehicle is fully driven by the user; no sensors are active in this operation mode.

Table 1. Driving modes of the vehicle

- User and educator interfaces. The whole system includes a user board as the I/O interface between the child and the system and an educator board to select the mode and the parameters of driving and the general control of the vehicle.

  The user board is placed opposite to the child by means of an articulated arm so that it can be placed and oriented accordingly. In its basic configuration, it comprises four buttons to command forward, backward, left, and right displacements and has an additional stop button. Every button includes a lighting pilot. Specific user interfaces have been allowed by our system. In particular, we provided single or dual button interfaces in combination with scanning approaches where the children were unable to cope with the standard interface.

  The educator board is comprised of a display and buttons used to configure the driving mode. The system is programmed so that five driving levels are available, from lower to higher child intervention (higher to lower vehicle autonomy).

  Besides of driving modes, the system allows for different driving speeds, driving time, minimum time between consecutive button pressings, ressing time, and scanning intervals.

- Obstacle system detection

  For the obstacle detection system, various technologies were evaluated, i.e., vision cameras, laser sensors, and IR sensors. Eventually, ultrasonic sensors were selected because of their relative simplicity (in terms of system configuration and signal processing capability) and for the intrinsic possibility of obtaining range information to obstacles and the possibility of configuring independent detection lobes without additional angular scanning.

  The detection principle is based on computing the time of flight between transmitted ultrasonic waves and received echoes, using specific algorithms for detecting the start

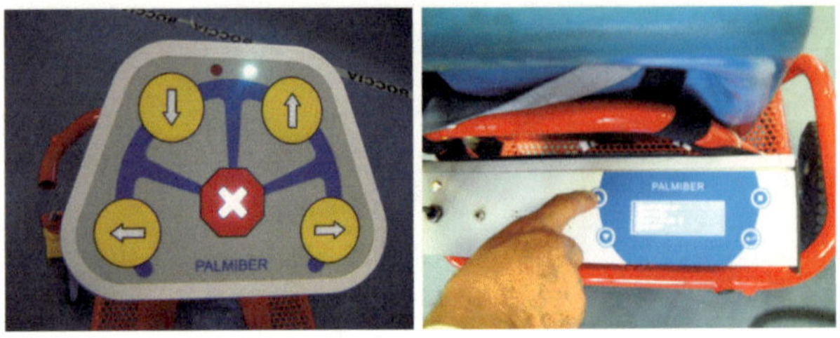

a) b)

Fig. 3. User and educator interfaces

of the echo signal. Concretely, a fast model-based algorithm for ultrasonic range measurements has been designed and evaluated an error estimation lower than 0.5 centimeter, Raya et al. (2008). The algorithm is based on the mathematical model of the ultrasonic envelope signal.

An architecture based on a digital signal processor (dsPIC) has been implemented to excite the ultrasonic transducer with a burst of 40 kHz composed of a pulse train. A set of seven ultrasonic sensors is placed around the vehicle. Each ultrasonic transducer works as both an emitter and a receiver. To reduce the effect of the dead zone (time period in which sensors cannot accurately detect the target) the number of exciting pulses is alternated between 5 and 15. Using 5 pulses, the system can detect obstacles close to the vehicle ($< 20cm$). Using 15 pulses, the system is capable to detect small obstacles such as chair legs.

3.3 Software tool for capturing and analysing the driving task

A distinctive feature of the PALMIBER system is the software tool to analyze the driving task. The vehicle capture the events during driving such as:

- Direction button pressed
- Direction proposed by the vehicle (voice message or lights)
- Time elapsed between above events (user's reaction time)
- Obstacle detection
- Driving path

All this information can be downloaded from the vehicle to the computer. The combination of these elemental events can be used to create an global evaluation function. For instance, if the user is suggested to reach a target destination, parameters such as path, time, manoeuvrability, reaction time can be captured. After several sessions, the learning progress can be evaluated. Fig. 4 depicts some screens of the software tool.

3.4 Clinical validation of the vehicle

The clinical validation of the vehicle was performed in the framework of the PALMA project, Ceres et al. (2005). In the experiments, a set of potential users was selected. The tests were carried out at Centro de Reabilitacão de Paralisia Cerebral Calouste Gulbenkian (CRPCCG) in Lisbon, Portugal. The criteria for selection included:

- Age range of 3 to 7 years, according to the recommendations for use of similar toy vehicles

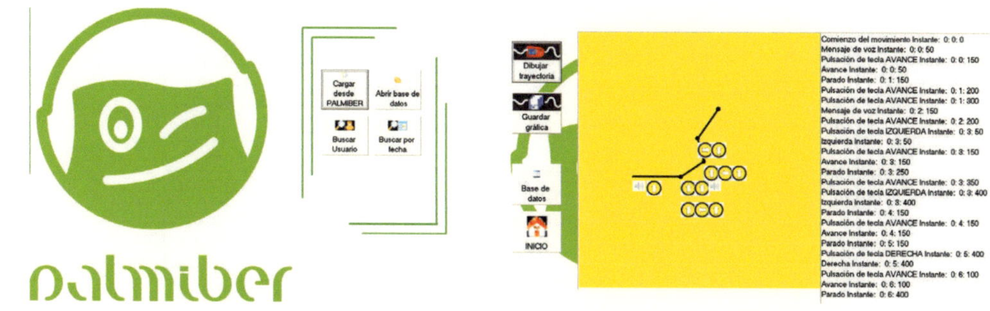

Fig. 4. Software tool for analyzing the driving task

- degree of mental retardation and difficulty in communication as well as mobility
- no previous experience of independent mobility

A total of 28 clinical trials were conducted on the various subjects. All clinical trials were recorded on video tape for subsequent through analysis. Each trial had a total duration of 15 minutes; 10 minutes were dedicated to free navigation around the classroom. This initial period was to allow the children to grow accustomed to the new navigation level so they would feel confident in driving the vehicle. Three minutes were dedicated to goal-oriented navigation (with increased difficulty according to the dexterity and navigation level). Targets were varied from passing through doors to delivering objects to friends in the classroom. Two additional free navigation minutes were allowed afterwards.

The analysis of results comprised both qualitative and quatitative parameters. Qualitative parameters (degree of stress and excitation, for example) were used to evaluate the attitude of children when using the rehabilitation tool. Quantitative parameters (number of uprising steps in the driving mode) give a direct measure of how the children increase their autonomy level in driving the vehicle.

The results showed that the learning process enabled the children to progress to more complex driving modes, even when the progress rate was different from user to user. The proposed method has a high impact on the cognitive development of the users, which together with its open and flexible structure makes PALMA an innovative and attractive rehabilitation tool.

3.5 Evaluation of the driving interfaces

The evaluation of the interaction between the child and the vehicle was performed in the framework of the PALMIBER project. The user board and a simple switch button were evaluated across the different driving modes. The tests were carried out at ASPACE Cantabria (Spain). Six children with CP were chosen. The inclusion criteria was:

- Age ranged between 18 months and 12 years old.
- Severe or moderate (only U1) motor disorder. Four limbs affected.
- No visual impairments

Table 3 collects the number and duration of the sessions. Table 2 shows the motor disorder profile of the users. Table 4 summarizes the proposed tasks according to each driving mode.

User	Motor disorder description
U1	Spastic and athetoid tetraplegia with extensor hypertonia. Bimanual grasp and moderate intellectual disability.
U2	Athetoid tetraplegia with flexor hypertonia. Mild intellectual disability.
U3	Spastic tetraplegia. Possible assisted manipulation requiring trunk fixation. Several attempts to reach objects. Severe intellectual disability.
U4	Spastic tetraplegia with global hypotonia. Grasping skills although requiring to be motivated. Severe intellectual disability.
U5	Mixed tetraplegia. Central spasticity and dystonia-athetosis in upper limbs. Severe intellectual disability.
U6	Spastic tetraplegia with global hypotonia. Lack of cervical and trunk control. Reaching and grasping with left hand. Severe intellectual disability.

Table 2. Motor disorder profile of the users

User	N^o sessions	Duration
U1	4	10-15 minutes/session
U2	4	10-15 minutes/session
U3	4	15-20 minutes/session
U4	9	15-20 minutes/session
U5	9	15-20 minutes/session
U6	12	10-15 minutes/session

Table 3. Number and duration of the sessions

Driving mode	Task description
Autonomous	First contact with the vehicle.
Cause-Effect	Driving across the room motivating the child to press any button
Training of the direction	Driving across the room motivating the child to press the suggested key (by voice and lights).
Decision of the direction	Driving the vehicle to pass between two obstacles

Table 4. Proposed tasks according to each driving mode.

- User U1. As expected, U1 presented the best results of interaction with the user board. This is coherent because his motor disorder has been characterized as moderate.

  He started at level 1 but he reached rapidly the level 2. His average reaction time was 3.5 seconds and success (press the suggested key) percentage was 100%. In the level 3, the user was capable to correct the trajectory to reach the two obstacles. Nevertheless, the corrections were influenced by the delay between action planning and execution. As conclusion the user board is an usable interface for U1.
- User U2. U2 started at level 1 (Cause-effect). The user held his hand on the user board. It facilitated the task because the reaching movement was not necessary. As result, the

reaction time was about 3 seconds with 92% of success. In level 2, the user had to press left and right buttons alternatively. In this case, the user could not hold his hand on the board because the task required to perform reaching movements. The average of the reaction time was 9 and 16 seconds for left and right respectively. As a consequence of motor disorders, the user could not reach more advanced driving modes.

- User U3. U3 started at level 0 and continued at level 1 (Cause-effect). U3 used a simple switch because according to the therapists, the direction board was not usable for him. The average reaction time was 9.6 seconds using the switch. The cognitive disability impeded to advance on driving modes.
- User U4. U4 started at level 0 and continued at level 1 (Cause-effect). At the beginning, U4 used the direction board and the success percentage was very low (65%). For this reason, the user start to use the simple switch. The reaction time and success percentage were 93% and 17.5 seconds using this interface. As result, the switch results more usable for this user. The cognitive disability impeded to advance on driving modes.
- User U5. The results for U5 were very similar to U4. U5 started at level 0 and continued at level 1 (Cause-effect). The success percentage was 63.5% with the direction board. The reaction time and success percentage using the simple switch was 13.6 seconds and 96% respectively. The cognitive disability impeded to advance on driving modes.
- User U6. The results for U6 were very similar to U4 and U5. U6 started at level 0 and continued at level 1 (Cause-effect). The success percentage was 52.67% with the direction board. The reaction time and success percentage using the simple switch was 9.2 seconds and 95% respectively. The cognitive disability impeded to advance on driving modes.

The findings shows that people with mild voluntary control of their upper limbs (U1) could drive the vehicle, whereas people with severe motor disorders present meaningful limitations. The simple switch is an alternative solution. In fact, this device is commonly used to access to other assistive devices such as computer. However, new technologies can be useful to create novel interfaces more adapted to user's needs.

A review of person-computer interfaces focused on people with cerebral palsy is presented in the following section. Our goal is to identify the shortcomnigs of the current person-computer interfaces and propose a novel solution to reduce them.

4. Human-computer interfaces focused on CP

Davies et al. (2010), published a systematic review on the development, use and effectiveness of devices and technologies that enable or enhance self-directed computer access by individuals with CP. The study showed that twenty-four studies had fewer than 10 participants with CP, with a wide age range of 5 to 77 years. International standards exist to evaluate effectiveness of non-keyboard devices, but only one group undertook this testing. Authors concluded that access solutions for individuals with CP are in the early stages of development.

The access solutions for individuals with CP can be divided into: 1) pointing devices, 2) keyboard modifications, 3) screen interface options, 4) Speech-recognition software and 5) Algorithms and filtering mechanisms.

Touch screens, Durfee & Billingsley (1999), switches with scanning approaches, Man & Wong (2007) and joysticks, Rao et al. (2000), are examples of pointing devices. The eye and face tracking systems have been widely researched in the last years, Betke et al. (2002); Mauri et al. (2006). Eye and gaze-based interfaces have the potential to be a very natural form of pointing, as people tend to look at the object they wish to interact with. Moreover, they do not require the user to wear any thing in contact with the body. In these approaches the system can track either the movement of the head, or the pupil's movement relative to the head. In the last case, the user's head must remain fixed in relation to the camera position. This is an critical aspect if the user has involuntary movements.

As regards the keyboard-based solutions, some studies have demonstrated that modifications improve speed and accuracy, Lin et al. (2008); McCormack (1990). Screen interface refers to scan through screen icons or change dinamically the icon position. Children with significant physical impairments (who are unable to point) use visual scanning and switches to select symbols. Symbol-prediction software is a method of access that involves highlighting a specific symbol within an array on the basis of an expected or predicted response, Stewart & Wilcock (2000). The prediction software reduces the response time required for participants but there is a trade-off between speed and accuracy.

Some devices are voice-based human computer interfaces in which a set of commands can be executed by the voice of the user. Speech-recognition software is difficult to customize for users with CP who have dysarthric speech. A combination of feedback information through auditory repeat and visual feedback may help users to reduce variability in dysarthric utterances and enable increased recognition by speech-recognition software, Havstam et al. (2003); Parker et al. (2006).

Algorithms and filtering mechanisms are focused on improving the accuracy of computer recognition of keyboard input or tracking of cursor movement. In connection with techniques to facilitate the cursor control, there are two different approaches: 1) target-aware and 2) target-agnostic. Target-aware techniques require the mouse cursor to know about, and respond to, the locations and dimensions of on-screen targets. Examples are gravity wells, Hwang et al. (2003), force fields, Ahlström & Leitner (2006) and bubble cursors, Grossman & Balakrishnan (2005). In contrast, few techniques are target-agnostic, meaning that the mouse cursor can remain ignorant of all on-screen targets, and targets themselves are not directly manipulated. Conventional pointer acceleration, Casiez et al. (2008), is by far the most common target-agnostic technique, one found in all modern commercial systems. Wobbrock et al. (2009), have designed an algorithm that adjusts the mouse gain based on the deviation of angles. However, the results showed general conclusions for all participants in which cerebral palsy was included among other disabilities such as parkinson or Friedreich's ataxia. Some mathematical analyses showed that additional modelling and filters within the computer software could theoretically improve icon selection when using a mouse as the input, but this was not tested in real time with participants, Olds et al. (2008).

Although there are many developments for disabled people in general, there are few evidences from motor disabled community using these alternative interfaces, Bates & Istance (2003). Moreover, most of the authors of access solutions affirm that the usability decreases dramatically when user suffers a severe motor disability. Our goal is to create an access solution that takes into account the particular limitations caused by the motor disorders. The first step will be to characterize the limitations of the user. The second step, will be to design filtering strategies to reduce the limitations.

5. Inertial human-computer interface for people with CP

5.1 The ENLAZA device

The novel interface proposed, called ENLAZA, is addressed to people with pathological movements, which involve voluntary and involuntary movements, such as tremor or spasms. The interface is based on inertial technology which will be useful to extract kinematic patterns of the voluntary and involuntary movements. Although all areas of the motor function are limited, limbs are usually more affected than the head motion in infants with CP, Wichers et al. (2009). Hence, the inertial interface ENLAZA is a head mounted device.

The interface consists of a headset with a commercial helmet and an inertial measurement unit (IMU), Fig. 5. The IMU (developed by the collaboration between the authors and Technaid S.L.) integrates a three-axis gyroscope, accelerometer and magnetometer. A rate gyroscope measures angular velocity by measuring capacitance and it is based on the Coriolis force principle during angular rate. The accelerometer measures the gravity and the acceleration caused by motions (by Hookes law). The magnetometer measures the earth magnetic field. The 3D IMU is based on MEMS technology and is available in a package measuring 27x35x13mm and its weight is 27 grams which is less than other sensors used in the field Rocon et al. (2005), Roetenberg et al. (2005). The 3D IMU is capable of sensing +/- 2.0 gauss, $+/-500^o/s$ angular rate and +/- 3g acceleration about three axes independently. It has an angular resolution of 0.05^o, a static accuracy less than one degree and a dynamic accuracy about 2° RMS.

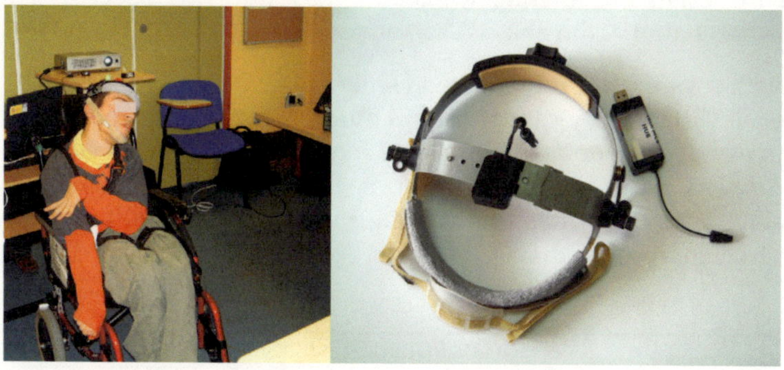

Fig. 5. Inertial interface ENLAZA

The interface ENLAZA translates the head rotations into cursor displacements. The Euler angles give the information about the rotation around the three axes (α, β and γ, rotations frontal, sagital and tranversal respectively). The calibration process gives the information to estimate the vertical and horizontal angular ranges (θ_v and θ_h). It consists of looking at three points on computer's screen (left-up, right-up and center). The Euler angles can be calculated following the next equation:

$$
\begin{aligned}
R_{GS} &= R_s \cdot (R_G)^{-1} \\
\alpha &= \arctan(-R_{GS}(2,3)/R_{GS}(3,3)) \\
\beta &= \arcsin(R_{GS}(1,3)) \\
\gamma &= \arctan(-R_{GS}(1,2)/R_{GS}(1,1))
\end{aligned}
\qquad (1)
$$

where R_G is the orientation matrix corresponding to center position and R_s is the orientation matrix at each frame. The pointer position can be calculated with the following equations:

$$\begin{aligned} x &= \gamma \cdot W/\theta_h \\ y &= \beta \cdot H/\theta_v \end{aligned} \tag{2}$$

where W and H are the screen width and height respectively.

The inertial interface was technically validated by five healthy users, Raya et al. (2010). The metric used was the Throughput defined by the ISO 9241-9 "Requirements for non-keyboard input device". The result was 2 bits/s which was found in agreement with other similar devices presented in the literature, Music et al. (2009).

An important distinction must be made respect to interfaces in the literature because the basis principle of the ENLAZA interface is to know the motor limitations of the user and adapt its performance to them. The device intends to be versatile and adaptable to different ranges of motion, postures and velocities. This aspect is considered to be essential because of the heterogeneous alterations caused by CP.

5.2 Characterization of cervical motor disorders using the inertial interface

The aim of the work presented in this section is to characterize the cervical motor disorders of people with CP using the inertial interface. The information extracted from these experiments will be used to design the filtering techniques to reduce the effect of the involuntary movements on the control of the ENLAZA interface.

Four people with severe CP were recruited (Table 5). Their mean age was 29 years (range: 26-35), Raya et al. (2011). They cannot control mouse pointer or keyboard. 3 healthy users participated to extract the normalized patterns for comparison. Tests with CP people were carried out in ASPACE Cantabria (Santander, Spain). ASPACE Cantabria has expertise in using some alternative devices as eye-tracking interfaces. Tests with healthy users were carried out in Bioengineering Lab CSIC (Madrid, Spain).

Subject	Motor Function Characteristics		
	Cervical tone	General tone	Associated movements
CP1	Extensor hypertonia	Extensor hypertonia	Athetosis
CP2	Dystonia	Distonia	Ballistics
CP3	Hypotonia	Hypotonía	No
CP4	Hypotonia	Dystonía	Dystonia

Table 5. Motor characteristics of participants with CP

The trial required participants to look at an on-screen target and dwell on it for selection. Then, the target changed its position following a sequential order. Participants were instructed to locate the cursor over the target as quickly as possible using head motion. There were 5 sessions during a week using an inertial pointing device. The kinematic data were recorded during the trial. The experiment consisted of reaching the target 15 times. This trial was repeated during 5 days one time per day. Therefore the target-reaching task was carried out 75 times in total. The target-reaching task is attractive because it provides a statistical description of the involuntary movements made during voluntary activity. The metrics to quantify the motor disorder were analyzed in three domains: 1) Time, 2) frequency and 3) space.

The analysis according to these three domains concluded the following statements:

- Kinematic analysis showed characteristic patterns for hypertonia and hypotonia disorders.
- Hypertonia (high muscle tone) causes involuntary movements at higher peak frequency (1.3Hz) than voluntary movements (0.3Hz).
- Hypotonia (low muscle tone) is characterized by abnormal postural activity. The frequency of hypotonic movements (0.3Hz-0.6Hz) is considerably similar to the frequency of voluntary motion.
- Voluntary and involuntary motion are combined at the same bandwitdh.
- The spatial-domain analysis showed that the unbalanced sagittal rotation was clearly wider respect to frontal and transverse, especially for hypotonic cases. It might be explained because the pull of gravity makes difficult to hold his head up.
- While hypertonia affects more to fine motor skills, hypotonia affects more to gross motor skills.
- Time-domain analysis revealed characteristic intention components with markedly increasing difficulties to control the pointer accurately around the target region.
- There was an association between motion frequency and capacity to maintain the trade-off between speed and accuracy.

Fig. 6 depicts the differences between voluntary and involuntary motor control in time domain. Fig. 7 compares the spectrogram of a healthy user with the spectrogram of a person with CP (CP1). Finally, Fig. 8 illustrates the range of motion for a healthy user and a person with cervical hypotonus (CP3) in which the head drops forward because of the gravity effect is evidenced.

6. Filtering algorithms to reduce the effect of the motor disorders

6.1 Joystick mode

Using the joystick mode can be a simple way to filter the involuntary movements. In this mode, pointer increases its position step by step according to the head pose. Fig. 9 illustrates this control mode. For instance, user looks at the grey area to stop the pointer. User looks at the red region to move the pointer towards the right direction.

Click task is easier using this mode because it is not necessary that user hold his/her head in an acurate position, only in a certain range.

6.2 Adaptive filters

The aim of the filtering techniques is to improve the target adquisition for users with CP. The conclusions described in the section 5.2 will be considered to design the adaptive filter. On one hand, filters based on separating frequency bands are not adequate because voluntary and involuntary movements are within the same bandwitdh. This fact eliminates simple low-pass filters.

Fitts' Law, Fitts (1954), models the voluntary motor behaviour in a reaching task. The reaching task consists of an initial movement that rapidly covers distance and a slower homing in

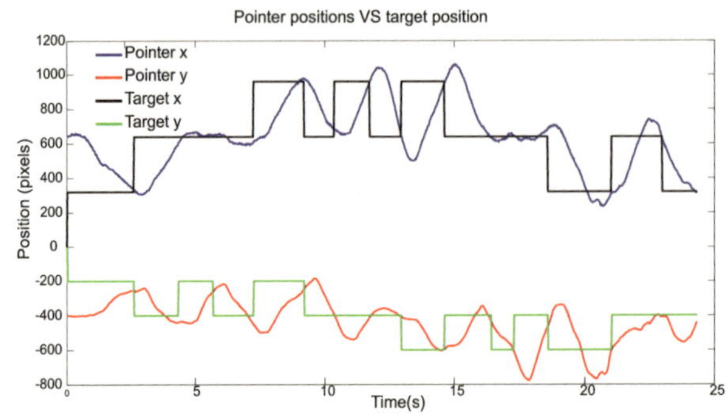

a)

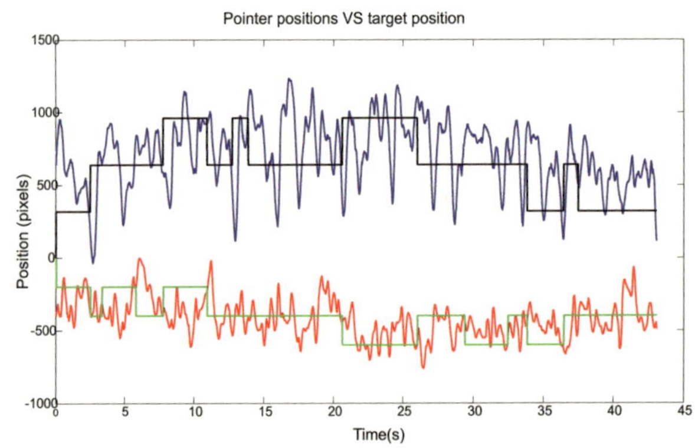

b)

Fig. 6. a) Pointer versus target positions (healthy subject), b) Pointer versus target positions (user with cervical hypertonia (CP1))

phase. The results showed that people with CP performed satisfactorily the initial movement with increasing submovements around the target region.

The proposed hypothesis states that long trajectories correspond to the initial movement (high probability of voluntary component) and rapid changes (generally performed around the target region) are undesirable. As consequence, the filter should have a dynamic gain adaptation related to the deviation of the cursor trajectory.

Adaptive filters are time-varying since their transfer function is continually adjusted driven by an reference signal that depends on the application. The general block diagram of adaptive filtering implementation consists of the prediction and update steps as depicted by figure 10.

The parameter k is the iteration number, $x(k)$ denotes the input signal, $y(k)$ is the output signal and $d(k)$ defines the desired signal. The error signal $e(k)$ is the difference between $d(k)$ and $y(k)$. The filter coefficients $W(k)$ are updated as stated by the error signal.

The equation parameters can be adjusted to track the movements of the mouse pointer. In some cases, the algorithms to track mouse positions assume a constant speed movement model. This assumption is reasonable since that sample period is very small compared with the movement speeds, Brookner (1998), i.e., the sample period adopted was $20ms$ and the

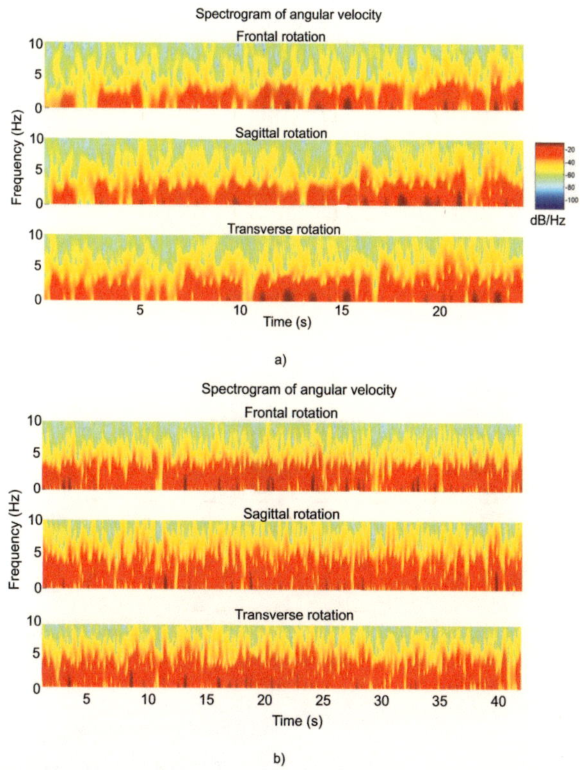

Fig. 7. a) Spectrogram of a healthy subject, b) Spectrogram of a person with cervical hypertonia (CP1)

voluntary movement estimated occurs in a bandwidth lower than 2Hz (75% of the power spectral density, Raya et al. (2011)).

The g-h filter (sometimes called α-β filter) is a simple recursive adaptive filter. It is used extensively as a tracking filter. The g-h algorithm consists of a set of update equations:

$$\dot{x}^*_{k,k} = \dot{x}^*_{k,k-1} + h_k \left(\frac{y_k - x^*_{k,k-1}}{T} \right) \tag{3}$$

$$x^*_{k,k} = x^*_{k,k-1} + g_k \left(y_k - x^*_{k,k-1} \right) \tag{4}$$

and prediction equations, Brookner (1998):

$$\dot{x}^*_{k+1,k} = \dot{x}^*_{k,k} \tag{5}$$

$$x^*_{k+1,k} = x^*_{k,k} + T\dot{x}^*_{k+1,k} \tag{6}$$

The tracking update equations or estimation equations (equations 3 and 4) provide the mouse pointer speed and position. The predicted position is an estimation of x_{k+1} based on past states and prediction, equations 5 and 6, and takes into account current measurement using updated states. T is the sample period and the parameters g_k and h_k are used to weight the measurements.

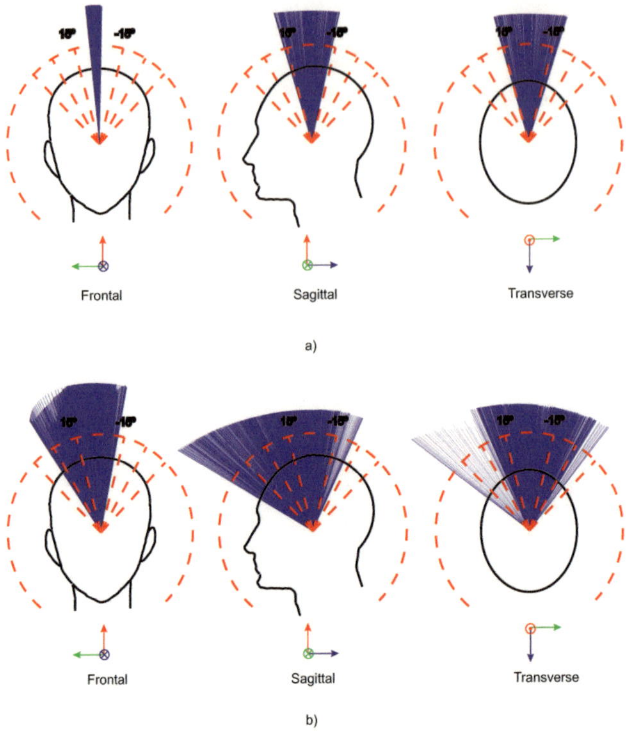

Fig. 8. a) ROM of a healthy subject, b) ROM of a person with cervical hypotonus (CP3)

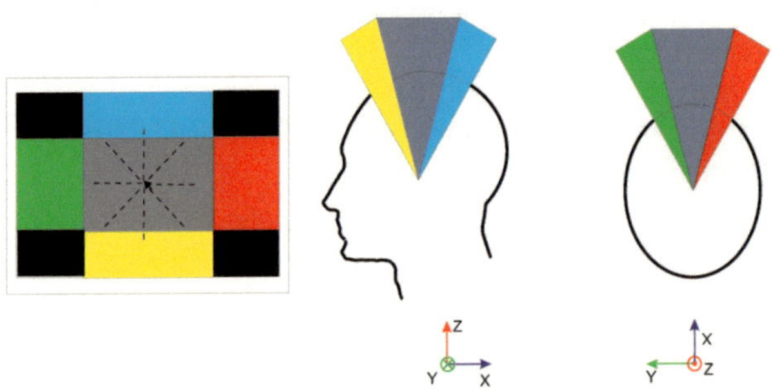

Fig. 9. Control space of joystick mode

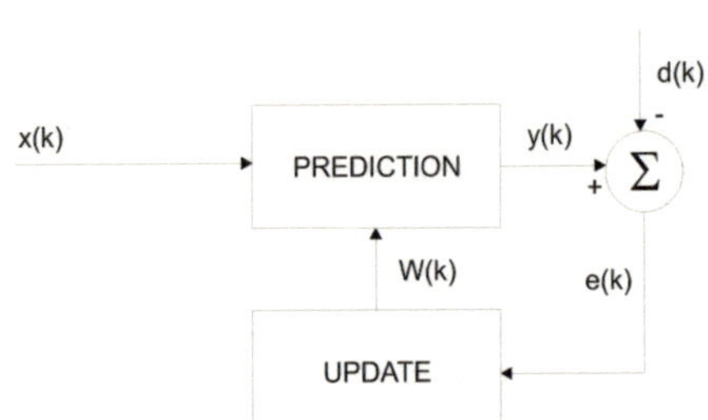

Fig. 10. Block diagram of adaptive filtering implementation

Benedict Bordner filter, Benedict & Bordner (1962), and Kalman filter are adaptive filters commonly used in tracking applications. These algorithms were successfully applied by some authors for tremor suppression, Gallego et al. (2010); Pons et al. (2007); Riviere & Thakor (1998.). The purpose of this investigation is to determine the feasibility of using these adaptive filters for reducing the motor disability effects caused by CP. In addition, we propose a *robust Kalman filter* that improves the performance of the classic kalman because it has been designed to detect outliers.

The performance of these algorithms was compared based on a kinematic descriptor called *segmentation*, McCrea & Eng (2005). The *segmentation* is a useful metric to estimate the corrections or submovements performed during the target-reaching task. Firstly, the function "remaining distance to the target versus time" is calculated. Secondly, the maxima of this function are calculated corresponding to movements performed in opposite direction relative to the target. Figure 11 illustrates an example of segmentation.

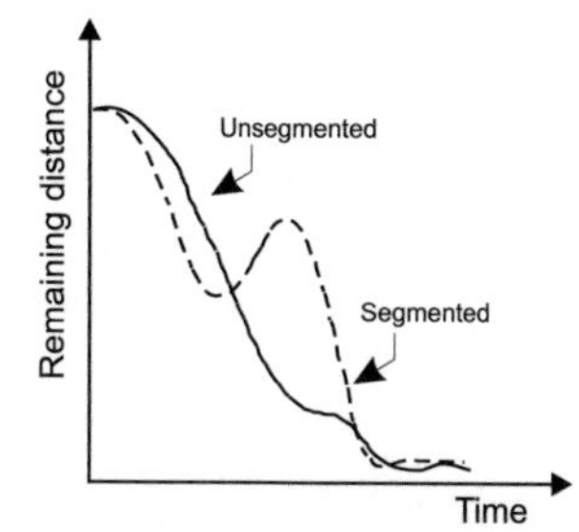

Fig. 11. Segmentation: Kinematic descriptor of the improvement introduced by the adaptive filter

6.3 Benedict-Bordner filter (BBF)

The Benedict-Bordner estimator is designed to minimize the transient error. Therefore, it responds faster to changes in movement speed and is slightly under damped, Bar-Shalom & Li (1998). The relation between filter parameters is defined by equation 7:

$$h = \frac{g^2}{2-g} \tag{7}$$

g-h gains are manually selected and static.

6.4 Kalman filter (KF)

The application of a Kalman filter requires that the dynamics of the target is represented as state space model, Kalman (1960). A simple kinematics approach based on the assumption of constant velocity process is suggested by some authors, , and is shown to track voluntary movements correctly. Figure 12 illustrates the block diagram of the Kalman filter.

The main difference respect to g-h filters is that a Kalman filter uses covariance noise models for states and observations. Using these, a time-dependent estimate of state covariance is updated automatically, and from this the Kalman gain matrix terms are calculated.

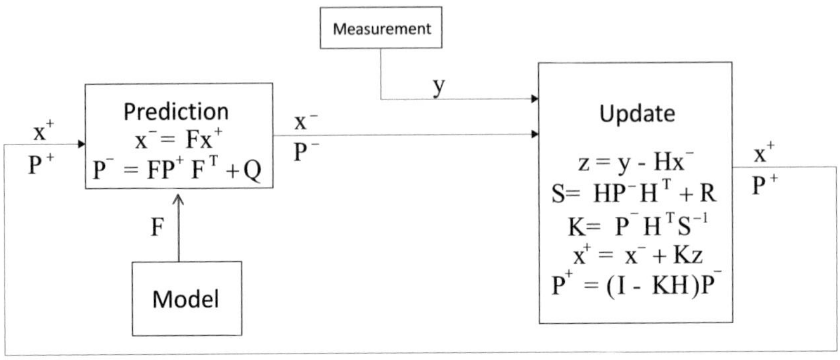

Fig. 12. Block diagram of the Kalman filter

6.5 Robust Kalman filter (RKF)

The Kalman filter is commonly used for real-time tracking, but it is not robust to outliers. The submovements around the target region caused by motor disorders can be considered as outliers respect to the constant velocity model (voluntary model). In our application, it is difficult to define a complete model of the pathological patterns because they are not repetitive. We propose to establish a model of voluntary model and consider the observations that lie outside the pattern of normal distribution as outliers. Using RKF, to the normal distribution containing the variances for the observations a second normal distribution is added which has larger variances than the first.

The robustification is based on the methodology of the M-estimators by following Huber's function, Huber (1981). The difference between the measurement and the estimation is weighted according to this Huber's function. For scalar observations, the Huber's function ψ_H of the form, Cipra & Romera (1991):

$$\varphi(Kz) = Kz \cdot min(1, b/|Kz|)$$

where $|Kz|$ is the euclidean norm. Figure 13 illustrates the Huber's function. Figure 14 depicts

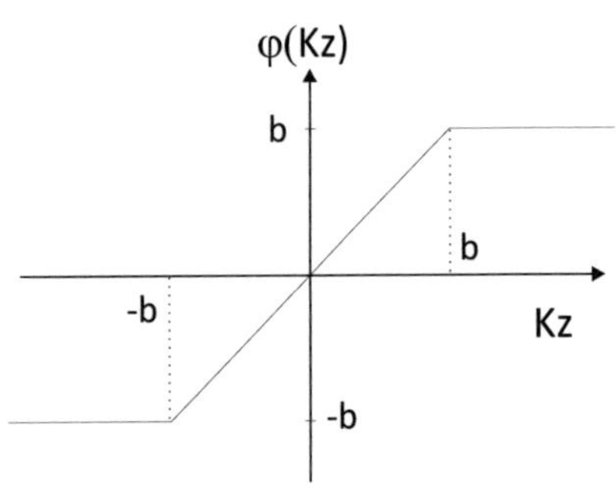

Fig. 13. Huber's function

the block diagram of the robust Kalman filter. The result is a very easy implementation and simple derivation of classic Kalman algorithm that includes the detection and elimination of undesirable data by an iterative downweighting of the outlying observations within the method of least squares.

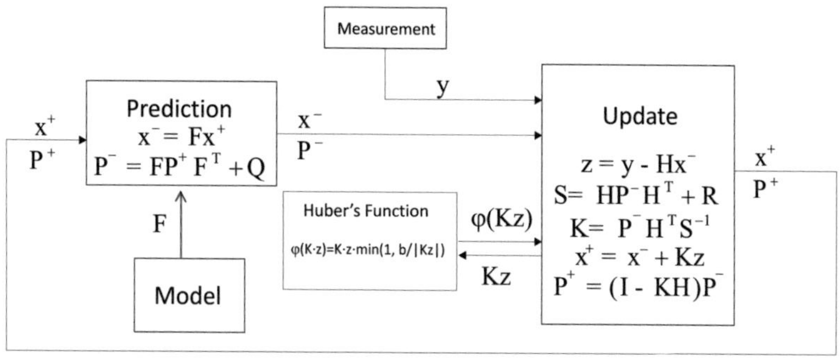

Fig. 14. Block diagram of the robust Kalman filter

6.6 Evaluation of the adaptive filters

The filtering algorithms BBF, KF and RKF were applied offline to the previously captured data. The kinematic data were registered during the experiments described in the section 5.2. The task consisted of reaching 15 targets which changed sequentially their location once reached (locate the pointer over the target). The task was repeated during five days. As result, a total of 75 reaching tasks were registered and post-processed.

The segmentation of the movement will give interesting information about the number of submovement. It is expected that the reduction in the number of submovements will improve the fine control being a critical aspect in the target adquisition for people with CP. Segmentation is estimated calculating the number of local maximums separated by $200ms$ of the function "Remaining distance versus time". The average of submovements was $M = 1.41(STD = 0.18)$ for healthy subjects. As expected, results showed higher number of submovement for participants with CP where the mean was about 8 submovements for CP1, CP2 and CP3 and slightly higher for CP4. Table 6 summarizes the number of submovements without and with filtering application. All filters considerably reduce the number of submovements reaching about a 65% of reduction.

User	Filtering algorithms %(Mean(std))			
	Without filter	BBF	KF	RKF
CP1	7.92(1.26)	4.83(0.96)	3.93(0.70)	3.5(0.77)
CP2	8.06(3.38)	3.97(1.71)	3.10(0.98)	2.83(0.85)
CP3	7.82(1.54)	4.08(1.22)	3.04(0.77)	2.77(0.76)
CP4	14.35(8.07)	7.02(4.09)	4.73(2.42)	4.64(2.39)

Table 6. Segmentation during the target-reaching task

The effect of the filtering techniques can be graphically shown. Figures 15b, 15c, 15d depict the target-reaching path without and with filtering for BBF, KF and RKF filters. Table 6 demonstrates that RKF filter had the best performance followed by KF and BBF filters.

BBF responds faster to changes in movement being able to filter movements of high frequency. However, in our application voluntary and involuntary movements can appear combined within the same bandwitdh. The detection and elimination of outliers (submovements) result useful for our application. The gain filter is modulated in real-time being lower during straight paths in which the prediction error is smaller. By means of outliers suppression and the dynamic filter gain, the initial movement that rapidly covers distance is smoothly filtered whereas the movements around the target are strongly filtered. As consequence, the fine control is facilitated.

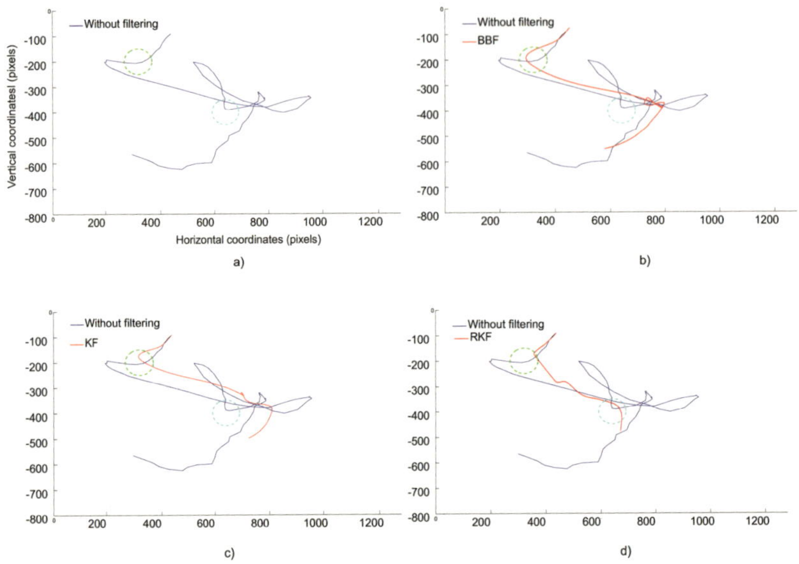

Fig. 15. Pointer path a) without filtering b) with BBF c) with KF d) with RKF

7. Functional evaluation of the ENLAZA interface as input device for computer

Once the filtering algorithm has been designed and validated technically, the ENLAZA interface and the filtering technique must be validated by people with CP. Three of the four people who participated in the experimentation described in section 5.2 participated. One of them, user CP2, could not participate because he was not in the centre ASPACE Cantabria when these trials took place. The table 5 summarized the motor profile of these users.

The task consisted in reaching 15 targets which changed their position once reached. One important difference respect to the experimentation described in section 5.2 was introduced: it is necessary to click on the target for the selection (not only crossing as in section 5.2). In this way, the reaching and the selection is evaluated, that means, the normal operation when using the computer. The condition for selection was to remain the pointer on a region of 60 pixels during 3.5 seconds. A pointing magnifier tool was used to facilitate the selection. The pointing magnifier has been developed by the AIM Research Group (University of Washington), Jansen et al. (2011). It was used by CP3 and CP4. CP1 had better results without it.

The metric used was the reaching time. The following three methods were compared:

- Target-reaching task without filter
- Target-reaching task with robust Kalman filter
- Target-reaching task using joystick mode

Table 7 summarizes the reaching time for each user and method. The main difference is observed in CP1. The adaptive filter RKF reduces about 10 times the reaching time.

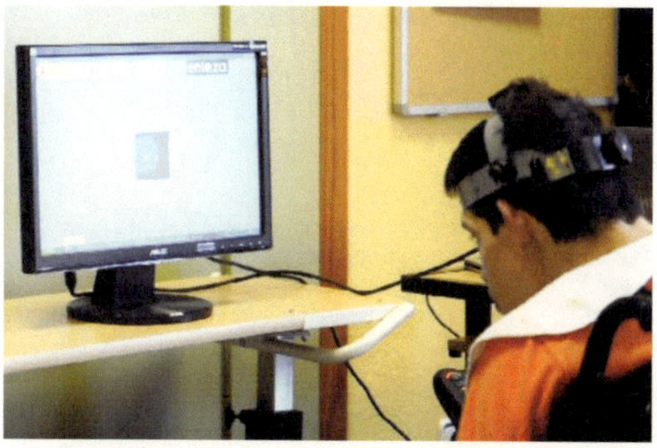

Fig. 16. Experiments with the ENLAZA interface at ASPACE Cantabria

As described in section 5.2 the fundamental frequency of CP1 is higher than voluntary movement. The filter reduces the effect of the movements of high frequency, so that, the fine control is improved. The reduction was 2 times for CP3 and CP4 (hypotonus cases). According to the results of section 5.2, the correlation with the voluntary motor control was higher for hypotonus cases. Therefore, the effect of filter is smaller. The joystick mode also facilitates the control. However, although the click task is facilitated, the reaching task can be considerably slower. RKF presents a better trade-off between reaching and selection. Fig. 16 depicts the a picture of the experiments at ASPACE Cantabria. In conclusion, the inertial

User	Methods (%(Mean(std)))		
	Without filter	Joystick mode	RKF
CP1	109(10.98)	15.67(11.70)	8.67(4.78)
CP3	44.16 (34.77)	19.23 (6.74)	18.08(14.82)
CP4	43.26 (37.30)	39.97 (21.26)	17.43 (12.20)

Table 7. Reaching time (seconds) for each user and method

interface ENLAZA is usable as input device for the computer. People with severe limitations to access to conventional interfaces such as mouse and keyboard could access to the computer with an average reaching time between 8 and 18 seconds. Robust Kalman filter facilitates the target adquisition reducing the effect of the involuntary movements on the control.

8. Functional evaluation of the ENLAZA interface as input device for driving the PALMIBER vehicle

As described in section 3.5 some children with adequate cognitive development could not advance on driving modes because of their physical disability. It was the case of U2. The aim of the work presented in this section is to evaluate how the ENLAZA interface can help to drive the vehicle PALMIBER.

The control mode was based on controlling the movement of the vehicle with the head's rotations. The first step consisted of a calibration process to set the reference. Later, the γ angle controlled the turns of the vehicle and the β angle controlled the forward and backward movements.

U2 participated in these experiments. U1 could not participate. U3-U6 did not participate because of their cognitive disability impeded the adquisition of spatial concepts.

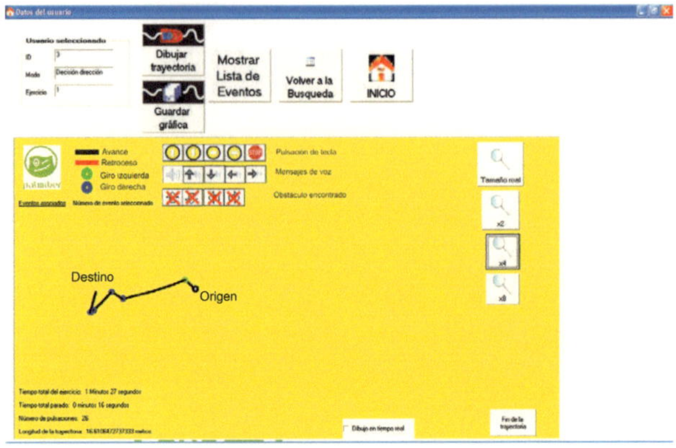

Fig. 17. Driving task analysis with the software tool

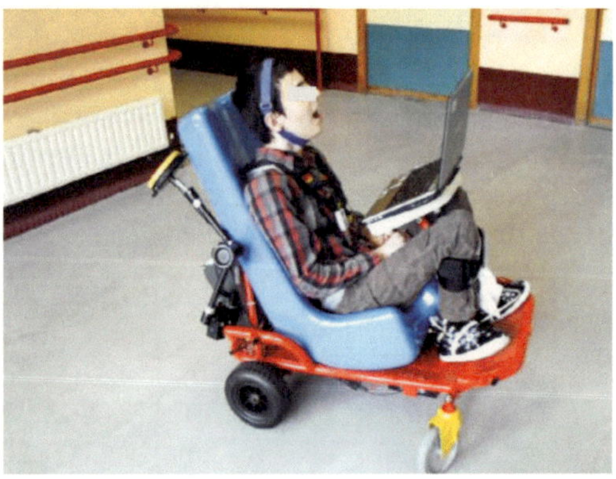

Fig. 18. Experiments of the ENLAZA as input device for driving the PALMIBER vehicle at ASPACE Cantabria

Nevertheless, U2 is a characteristic case because he knows the spatial concepts but his physical disability limit the capacity to drive the vehicle.

U2 started at the level 1 (Cause-effect). The reaction time was about 3 seconds being very similar to the result with the user board. Following, the driving mode 2 (left-right) was proposed. The average time was 4.2 seconds for both directions. This is a meaningful result compared to the reaction time using the user board (9 and 16 seconds, left and right directions respectively).

Finally, U2 was proposed to perform a functional task consisted in reaching a target destination (level 3). Fig. 17 illustrates an example of the path executed to reach the point. Although, some deviations respect to the ideal path exist, the user reached the goal in a relatively short time. The result is meaningful because U2 could not select the directions with the keys of the user board. Additionally, the control of the directions using the head is more intuitive than pressing button with arrow symbols. This user cannot control a conventional joystick, for instance to control his wheelchair. Therefore, it is expected that the inertial interface can control other types of assistive devices. Fig. 18 depicts a picture of the experiments at ASPACE Cantabria.

9. Conclusions and future work

The aim of this work was to study, design and validate new strategies for mobility and interaction of people with CP. A review of the mobility devices for CP showed that there are a few devices specifically focused on the integral development of the child. A robotic vehicle to promote the integral development of children with CP has been presented. The vehicle design is versatile and open to adapt to the wide variability of motor and cognitive alterations caused by CP.

Generally, human-machine interaction is a critical aspect in people with CP. Recently, many research groups are studying new channels of interaction. There are different advanced solutions for people who cannot use conventional interfaces. However, most of the authors affirm that the usability decreases dramatically when user has a severe motor disability. We proposed a new interface based on inertial technology. The inertial interface allows to characterize the pathological motion of the user. This characterization is an useful information to know the skills and limitations of users in order to create strategies to help them.

A new filtering algorithm (robust Kalman filter) has been designed and validated to reduce the effect of the involuntary movements on the control of the device. Using this approach, the submovements around the target region are reduced and the fine motor control is facilitated. The inertial interface was validated as input device for the computer by users with CP.

Finally, the inertial interface was validated as input device for driving the vehicle. Using this device, the delay between the action planning and its execution is reduced. As result, the user can control the direction of the vehicle and correct the path in the desired instant. Additionally, the relation between movement direction and the symbol on the board requires abstraction which can result difficult for these users. The head's rotation is an more intuitive method of control.

The future work will be focused on:

- increasing the number of subjects
- testing the interfaces with people with other disabilities (e.g. stroke, spinal cord injury)
- testing the filtering algorithm with other interfaces (e.g. eye or face tracking)
- testing the interface with real software applications
- evaluating the learning progress of children on the use of the vehicle in long term

10. Acknowledgment

Authors thank ASPACE Cantabria, especially to Teresa González and Antonio Ruiz and professionals and people who took part in this work. This work was supported by PALMIBER project (IBEROEKA), ENLAZA project (IMSERSO) and HYPER project (CONSOLIDER-INGENIO 2010, Spain).

11. References

Ahlström, D., H. M. & Leitner, G. (2006). An evaluation of sticky and force enhanced targets in multitarget situations., *Proc. NordiCHI '06. New York: ACM Press, 58-67.*

Azevedo, L. (2006). *A Model Based Approach to Provide Augmentative Mobility to Severely Disabled Children through Assistive Technology*, PhD thesis, Universidad del País Vasco.

Bar-Shalom, Y. & Li, X. (1998). *Estimation and Tracking: Principles, Techniques, and Software.*, Artech House Publishers.
Bates, R. & Istance, H. (2003). Why are eyemice unpopular?adetailed comparison of head and eye controlled assistive technology pointing devices, *Univ Access Inf Soc* 2: 280–290.
Bax, M. (1964). Terminology and classification of cerebral palsy, *Developmental Medicine and Child Neurology* 6: 295–297.
Benedict, T. & Bordner, G. (1962). Synthesis of an optimal set of radar track-while scan smoothing equations, *IRE Transactions on Automatic Control* 7(4): 27–32.
Betke, M., Gips, J. & Fleming, P. (2002). The camera mouse: visual tracking of body features to provide computer access for people with severe disabilities, . *IEEE Transactions on Neural Systems and Rehabilitation Engineering* 10, No.1: 1–10.
Borggraefe, I.and Schaefer, J. S., Klaiber, M., Dabrowski, E., Ammann-Reiffer, C., Knecht, B., Berweck, S., Heinen, F. & Meyer-Heim, A. (2009). Robotic-assisted treadmill therapy improves walking and standing performance in children and adolescents with cerebral palsy, *XVIII Annual Meeting of the European Society of Movement Analysis for Adults and Children (ESMAC) 14-19 September*.
Bourhis, G. & Agostini, Y. (1998). The vahm robotized wheelchair: System architecture and human man-machine interaction, *J. Intelligent Robot. Syst.* 22(1): 39–50.
Brookner, E. (1998). *Tracking and Kalman Filtering Made Easy.*
Cans, C. (2000). Surveillance of cerebral palsy in europe: a collaboration of cerebral palsy surveys and registers, *Developmental Medicine Child Neurology* 42: 816 824.
Casiez, G.and Vogel, D., Balakrishnan, R. & Cockburn, A. (2008). The impact of control-display gain on user performance in pointing tasks., *Human-Computer Interaction* 23 (3): 215–250.
Ceres, R., Pons, J., Calderón, L., Jiménez, A. & Azevedo, L. (2005). Palma, a robotic vehicle for assisted mobility of disabled children, *IEEE Engineering in Medicine and Biology magazine* 24,No. 6: 55–63.
Cipra, T. & Romera, R. (1991). Robust kalman filter and its application in time series analysis, *Kybernetika* 27(6): 481–494.
Civit, A. (1998). Tetranauta: A wheelchair controller for users with very severe mobility restrictions, *roc. 3rd. TIDE Congreso and Helsinki and Findland andpp336-341*.
Davies, C., Mudge, S., Ameratunga, S. & Stott, S. (2010). Enabling self-directed computer use for individuals with cerebral palsy: a systematic review of assistive devices and technologies, *Developmental medicine & Child Neurology* 52(6): 510(6).
Durfee, J. & Billingsley, F. (1999). A comparison of two computer input devices for uppercase letter matching., *Am J Occup Ther* 5: 214–20.
Fitts, P. (1954). The information capacity of the human motor system in controlling the amplitude of movement., *Journal of Experimental Psychology* 47(6): 381–391.
Gallego, J. A., Rocon, E., Roa, J., Moreno, J. C. & Pons, J. L. (2010). Real-time estimation of pathological tremor parameters from gyroscope data, *Sensors* 10(3): 2129–2149.
Galloway, J. C., Ryu, J.-C. & Agrawal, S. K. (2008). Babies driving robots: self-generated mobility in very young infants, *Intelligent Service Robotics* 1 (2): 123–134.
Gesell, A. (1966). *The First Five Years of Life.*, Metheun.
Grossman, T. & Balakrishnan, R. (2005). The bubble cursor: Enhancing target acquisition by dynamic resizing of the cursor's activation area., *Proc. CHI '05. New York: ACM Press, 281-290.*

Havstam, C., Buchholz, M. & Hartelius, L. (2003). Speech recognition and dysarthria: a single subject study of two individuals with profound impairment of speech and motor control., *Logoped Phoniatr Vocol* 28: 81–90.

Hoyer, H., Borgolte, U. & Hoelper, R. (1997). An omnidirectional wheelchair with enhanced safety comforts, *Proc. International Conference On Rehabilitations Robotics, University of Bath.*

Huber, P. (1981). *Robust Statistics.*

Hummel, F. C. & Cohen, L. G. (2005). Drivers of brain plasticity, *Curr Opin Neurol* 18: 667–74.

Hwang, F., Keates, S., Langdon, P. & Clarkson, P. (2003). Multiple haptic targets for motion-impaired computer users, *Proc. CHI '03. New York: ACM Press, 41-48.*

Jansen, A., Findlater, L. & Wobbrock, J. O. (2011). *ACM CHI Conference on Human Factors in Computing Systems.*

Johnson, A. (2002). Prevalence and characteristics of children with cerebral palsy in europe, *Developmental Medicine and Child Neurology* 44,No. 9: 633.

Kalman, R. (1960). A new approach to linear filtering and prediction problems., *Journal of Basic Engineering - Transactions of the ASME* 82: 35–45.

Katevas, N., Sgouros, N., Tzafestas, S., Papakonstantinou, G., Bishop, P. B. J., Tsanakas, P. & Koutsouris, D. (1997). The autonomous mobile robot senario: A sensor-aided intelligent navigation system for powered wheelchairs, *IEEE Robot. Automat. Mag.* 14: 60–70.

Lin, Y., Chen, M., Yeh, C., Yeh, Y. & Wang, H. (2008). Assisting an adolescent with cerebral palsy to entry text by using the chorded keyboard., *In: Miesenberger K, Klaus J, Zagler W, Karshmer A, editors. Computers helping people with special needs: 11th International Conference, ICCHP 2008, Linz, Austria, July 9-11, 2008 - Proceedings. Heidelberg, Germany: Springer-Verlag: 1177-83.*

LoPresti, E. F., Mihailidis, A. & Kirsch, N. (2004). Assistive technology for cognitive rehabilitation: State of the art., *NEUROPSYCHOLOGICAL REHABILITATION* 14 (1/2): 5Ű39.

Lynch, A., J-C, R., Agrawal, S. & J., G. (2009). Power mobility training for a 7-month-old infant with spina bifida, *Pediatric Physical Therapy* 21(4): 362–368.

Man, D. W. K. & Wong, M. L. (2007). Evaluation of computer-access solutions for students with quadriplegic athetoid cerebral palsy., *American Journal of Occupational Therapy* 61: 355–364.

Mauri, C., Granollers, T., Lorés, J. & García, M. (2006). Computer vision interaction for people with severe movement restrictions., *An Interdisciplinary Journal on Humans in ICT Environments.* 2 (1): 38–54.

Mazo, M., García, J., Rodríguez, F., Ureña, J., Lázaro, J., Espinosa, F. & the SIAMO research team. (2002). Experiences in assisted mobility: the siamo project., *Proceedings of the 2002 IEEE International Conference on Control Applications.*

McCormack, D. (1990). The effects of keyguard use and pelvic positioning on typing speed and accuracy in a boy with cerebral palsy., *Am J Occup Ther* 44: 312–5.

McCrea, P. & Eng, J. (2005). Consequences of increased neuromotor noise for reaching movements in persons with stroke, *Exp. Brain Res* 162: 70–77.

Music, J., Cecic, M. & Bonkovic, M. (2009). Testing inertial sensor performance as hands-free human-computer interface, *WSEAS TRANSACTIONS on COMPUTERS.*

Olds, K., Sibenaller, S., Cooper, R., Ding, D. & Riviere, C. (2008). Target prediction for icon clicking by athetoid persons, *Proceedings-IEEE Int Conf Robot Autom 19-23 May 2008, article no. 4543507: 2043-8.*

Parker, M., Cunningham, S., Enderby, P., Hawley, M. & Green, P. (2006). Automatic speech recognition and training for severely dysarthric users of assistive technology: the stardust project., *Clin Linguist Phon* 20: 149–56.

Pons, J., Rocon, E., Ruiz, A. & Moreno, J. (2007). *Upper-Limb Robotic Rehabilitation Exoskeleton: Tremor Suppression*, Itech Education and Publishing 978-3-902613-04-2, chapter 25, p. 648.

Rao, R., Seliktar, R. & Rahman, T. (2000). Evaluation of an isometric and a position joystick in a target acquisition task for individuals with cerebral palsy., *IEEE Trans Rehabil Eng* 8: 118–25.

Raya, R., Frizera, A., Ceres, R., Calderón, L. & Rocon, E. (2008). Design and evaluation of a fast model-based algorithm for ultrasonic range measurements, *Sensors and Actuators A: Physical* 148: 335–341.

Raya, R., Roa, J., Rocon, E., Ceres, R. & Pons, J. (2010). Wearable inertial mouse for children with physical and cognitive impairments, *Sensors & Actuators: Physical* 162: 248–259.

Raya, R., Rocon, E., Ceres, R., Harlaar, J. & Geytenbeek, J. (2011). Characterizing head motor disorders to create novel interfaces for people with cerebral palsy, *International Conference on Rehabilitation Robotics*.

Riviere, C. & Thakor, N. (1998.). Modeling and canceling tremor in human-machine interfaces., *IEEE Engineering in Medicine and Biology* 1: 29Ű36.

Rocon, E., Ruiz, A. & Pons, J. (2005). On the use of rate gyroscopes for tremor sensing in the human upper limb, *of the International Conference Eurosensors XIX*.

Roetenberg, D., Luinge, H. J., Baten, C. T. M. & Veltink, P. H. (2005). Compensation of magnetic disturbances improves inertial and magnetic sensing of human body segment orientation., *Ieee transactions on neural systems and rehabilitation engineering* 13(3): 395–405.

Rofer, T. & Lankenau, A. (1998). Architecture and applications of the bremen autonomous wheelchair, *Proc. 4th Joint Conf. Information Syst.* 1: 365–368.

Stewart, H. & Wilcock, A. (2000). Improving the communication rate for symbol based, scanning voice output device users., *Technol Disabil* 13: 141–50.

Wichers, M., Hilberink, S., Roebroeck, M., Nieuwenhuizen, O. V. & Stam., H. (2009). Motor impairments and activity limitations in children with spastic cerebral palsy: a dutch population-based study., *Journal of Rehabilitation Medicine* 41(5): 367–374 (8).

Winter, S., Autry, A. & Yeargin-Allsopp, M. (2002). Trends in the prevalence of cerebral palsy in a population-based study, *Pediatric* 110,No. 6: 1220–1225.

Wobbrock, J., Fogarty, J., Liu, S., Kimuro, S. & Harada, S. (2009). The angle mouse: Target-agnostic dynamic gain adjustment based on angular deviation, *Computer Human Interaction Conference*.

9

Brazilian Assistive Technology in Bath or Shower Activity for Individuals with Physical Disability

Fabiola Canal Merlin Dutra[1],
Alessandra Cavalcanti de Albuquerque e Souza[2],
Cláudia Regina Cabral Galvão[3],
Valéria Sousa de Andrade[2], Daniel Marinho Cezar da Cruz[4],
Daniel Gustavo de Sousa Carleto[1,2] and Letícia Zanetti Marchi Altafim[3]
[1]*Empresa Cavenaghi - São Paulo/SP*
[2]*Universidade Federal do Triângulo Mineiro, Uberaba/MG*
[3]*Universidade Federal da Paraíba, João Pessoa/PB*
[4]*Universidade Federal de São Carlos - São Carlos/SP*
Brazil

1. Introduction

The current market tends to expand products designed to serve large portions of the population, providing tools, objects, and furnishings with a design that meets the requirements or needs of people with or without disabilities. Regarding that perspective there has been an increase in the variability of these products marketed to the most diverse purposes, including for daily life activities.

This chapter is the result of a research about assistive technology devices available in the Brazilian market for positioning handicapped people during bath or shower, activity that is routinely performed by people. This activity is directed to the care of the person with his own body and is known as a basic activity of daily living, approaching particular demands, characteristic objects and specific contexts of each person (American Association of Occupational Therapy [AOTA], 2008).

The variation that a physical disability may cause is huge. It might range from the absence of one of the lower limbs or to its paralysis, to the total absence of postural control (characterized by a deficit or absence of cervical control - head, neck and trunk), leading these individuals to need external assistance to support his participation through their involvement in the bath or shower, especially in while seated.

Besides, the variability of the disability may require specific characteristics in the selected equipment for positioning the person during the bath or shower activity. These specific characteristics are materialized in accessories individualized as belts, head supports, lateral seatbelts for trunk, that will promote stability, security and comfort to individuals with disabilities and provide adequate posture to the caregiver/assistant, when necessary.

2. Bathing or showering

Personal hygiene is the key factor in the life. Bathing or showering may be described as a sum of experiences that provide maintenance of health, well being and relaxation (Gooch, 2003).

The action of bathing or showering as well as lathering, rinsing and drying parts of the body, approaches the possibility of the person to maintain an adequate positioning while performing bath or shower activity, transferring from a surface to another and to possible bath or shower positions (AOTA, 2008).

Gooch (2003) considers bath or shower as a complex and potentially dangerous activity to all people since it involves a combination of water, foam and a smooth surface, characterized by a tub surface of the floor. Considering this context and the disabled or handicapped person, such situation may become even more complex.

Bathing or showering a person with severely impaired motor function is an activity that demands planning, care and overestimated attention with safety, besides increasing elaboration of complexity and demand degree that come with his limitations and restrictions.

When taking into account bath or shower in relation to children, their growth must be considered while selecting and adjusting the chosen product for the sitting positioning during bath or shower activity. An inadequate choice can result, in long term, in a not pleasurable activity for the child and his caregiver (Tabaquim & Lamônica, 2004). Comfort and safety are also essential characteristics in the process to select the equipment to maintain the adequate sitting posture during bath or shower task of a disabled person.

Bath or shower independence is not existent in people with physical limitation or decreased strength, decreased range of motion (ROM), low or lost postural stability and limited or small ability to manipulate (dexterity).

Thus, if a person remains unstable while seated, he certainly is going to have difficulty to relax and stay on that position. Upper limbs commitment, characterized by weakness and movement limitation, may make complicate to run such posture adjustment and maintenance. In addition, rarely can this situation be associated to osteoarticular deformities caused by contractures, mainly on the knees and hips, which contributes even more to an unstable sit (Shephered, 2001).

A more adequate and appropriate solution involves a selection and denotation of a bath or shower equipment that must necessarily follow a "*hierarchical path*" to be prescribed, in which it should be primarily thought from the simplest adaptations till a specific bath/shower, being that in more complex cases, environment changes may also become necessary (Gooch, 2003).

The solution in solving such problem requires an intervention with compensatory strategies such as by using assistive technology devices, which comprises interfering on the context - that is, on the environment in which bath or shower task will be carried out - or on the demand of the task with adapted tools and utensils and with adapted furnished appropriate to the maintenance of the person in the sitting posture (e.g., a bath chair).

Assistive technology (AT), a science that combines various professions, such as health professionals (physical therapists, occupational therapists, speech therapist, and so forth), from technology area (engineers, architects, among others) and other areas, is defined in Brazil by the Technical Help Committee (Comitê de Ajudas Técnicas [CAT], 2001) as:

'an area of knowledge, interdisciplinary characteristic, the includes products, resources, methodologies, strategies, practices and services that aim to promote function, related to activity and participation, of people with disability or reduced mobility, aiming at autonomy, independence, quality of life and social inclusion'.

Consequently, AT is, thus, any technology developed to permit the increase of autonomy and independence of disabled people or the ones with reduced independence in their domestic or daily activities of daily living (BRAZIL/Ministério MCT, 2007).

In this context, AT devices are purposed to aid, enhance or promote the person's involvement on the task to be performed - that is, shower or bath -, putting satisfaction and occupational performance increment forward or, with respect to changes in physical context, the application of AT on the environment promoting environmental compliance.

Therefore, the suitable sitting positioning of a handicapped person while performing the bath or shower activity may be facilitated by selecting an AT device adjusted to the person's need or by compensating the physical contexts.

According to Shephered (2001), people who have problems to maintain their posture and movements, frequently lose control to assume or keep a stable posture during performance activities and are benefited in relation to an adaptive positioning.

Choosing an assistance device, that is, any equipment that promotes the increase and/or the participation of a person with disability on the execution of an activity may be a complex task. All limitations and needs of the user, caregiver and environment must be considered since their sum will define how the device will function and how its function will be perceived (Pain & McLellan, 2003).

Pain and McLellan (2003, p. 396) report in their study that for the indication of a device it must be taken into account that "there is a complex interaction between different factors, such as the user and the caregiver needs, the product characteristics and the environment that will be used". In order to guide the selection of such device, mainly when dealing with people with more complex needs, it is important that the individual clients' priorities and needs are clear.

The success of using and accepting assistive devices by a users and/or their caregivers, whether if they are produced in series (in mass) or tailor made, frequently involves a sum of professionals and factors, such as economic, ergonomic and aesthetic, in their development phase. Together, they add values to the product that ultimately is intended to expand or facilitate people and/or caregivers participation in the activity, reflecting their improved quality of life.

3. Equipments

Exploratory analysis of AT products, designed to meet the demands imposed by handicapped people in their different needs, resulted in the exposure of various types of equipments aimed to bath or shower. Most of these equipments have specific physical characteristics in order to attend meet the particular needs imposed by that user.

Among the products designed to bath or shower, the available equipments range from a simple system of seat (stool or regular chair) to specific equipments to people with serious motor commitment, wheelchair users and unable to remain seat without full support during the bath or shower activity.

Currently, the group of assistive devices for bathing or showering is available in Brazil in specialized stores for medical products, catalogues that show assistive technology products and internet websites, varying in their characteristics (Santana et al., 2009).

4. Stool/chair

The more commonly available equipments to help a person who needs support to remain sitting during shower or bath because of temporary or permanent motor commitment. There are many models in the market which can be made specifically for this purpose, as well as other ones that are available for other purposes but adapted for such task. Most of the conventional models are made of plastic with rubberized coating with variations in their design in relation to the model and dimension of the seat and backrest, armrest and variation in height (Figure 1 and 2).

(A) (B)

Fig. 1. Models of stools: (A) Four support stool with height adjustment; (B) Edge stool with adjustable height.

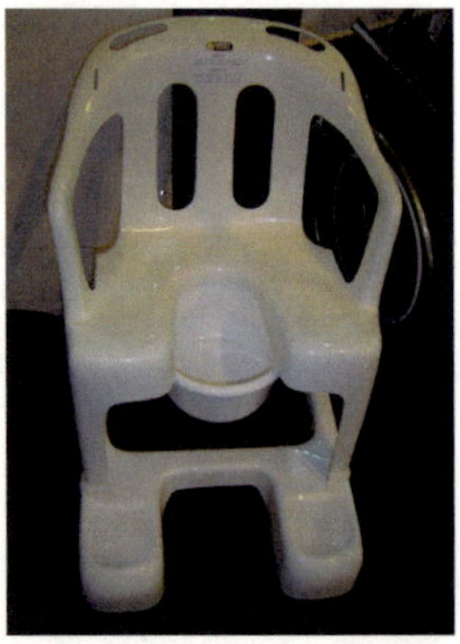

Fig. 2. Chair with cut out seat.

Despite the variation in available models of stools and chairs, some aspects deserve attention when choosing the equipment. The seat needs to be in accordance with the dimensions of the user in order to provide comfort and safety. Besides it is important to note their functional aspect to make bath or shower easy, like observing if there are holes that allow better flow drainage during the course of activity, if they show a front cut for making personal hygiene easier, if the seat is removable, what facilitates cleaning and maintenance of the equipment, and if it offers height adjustment promoting feet support on the ground.

Another factor that determines the type of the equipment is related to the base (feet) of the device, in which the model may vary between four supports (Figure 1A) in a three support model (Figure 1B) and type of edge, whose measure may vary according to the model. It is important to point out that, in order to keep the user safety, it is important that the stool/chair has rubber tips or suction cups on the feet that are connected to the ground to ensure stability in the environment. In addition, there are models that offer as an accessory the armrest, which increase the stability of the person in the sitting posture. After all, considering all these aspects for a referral/prescription is very important to ensure a better match between the equipment and the user.

5. Bath chair

Another type of AT equipment for showering or bathing is the wheelchair that can be attached to the toilet. This type of chair is usually light, manageable, may have padded seat and backrest made of fabric and armrests that can be articulated to make transferring easy to the chair. Some models may exhibit different accessories accomplished to the chair, as toilet paper support and the possibility of removing parts for cleaning.

Figure 3 (A and B) presents the most commonly used models of bath chair that are found in the Brazilian market, which are made available by reference companies for manufacturing and marketing of AT products. These models have similar physical characteristics, much of it designed to meet the requirement imposed by the adult audience, and is usually related to their design and dimensions. Typically, these bath chair models have their structures made of steel or aluminum, a seat toilet attached to the metal support structure and wheels different in the ring dimension and brakes. Besides, other specific characteristics may differentiate the product lines, as the present of removable armrests, fixed footrests (Figure 3A) or movable footrests, retractable armrest, removable footrest, padded backrest, beyond the ability to be folded for easy transport and storage (Figure 3B).

Usually these models of AT devices are suitable for people with mild or moderate physical disability or to the ones who present reduced mobility and need support to remain seated during bathing or showering, but are able to control body parts while seated, like cervical control and total or partial trunk control. This is because their structures are provided with a simple seat and back system that make it difficult to maintain proper sitting posture of people with severe motor impairment. Moreover, the options for the equipment dimensions end up restricting their indication for adult audiences.

Figure 4 represents another bath chair/hygiene chair model that represents a seat of characteristics designed to meet specific needs of people who have a more severely compromised motor capacity. The presence of removable head support, with possibility of height and depth adjustment, is considered an essential item when it cannot maintain keep

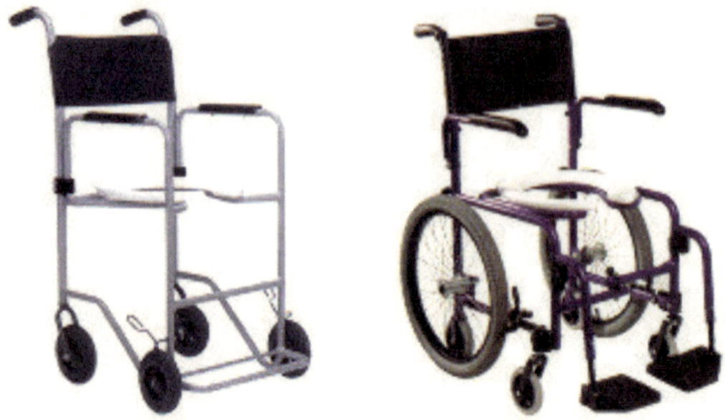

(A) (B)

Fig. 3. Adult aluminum shower bath chairs: (A) Fixed structure and (B) Folding structure for closing.

the position of the head against gravity; moreover, the bending system of the seat and back group ("tilt") of up to 35º allows a better accommodation for a person with absent or insufficient trunk control, who are unable to remain seated without support.

The presence of arms and legs support, which might be retractable or removable, supplement the necessary resources that permit a better accommodation of the user. Although the equipment presents more adequate features, the needs of more severely compromised people in the physical aspect, and the possibility of changing the adult toilet seat to a child one, implies that the dimensions offered by the equipment makes its designation more appropriate especially to adults.

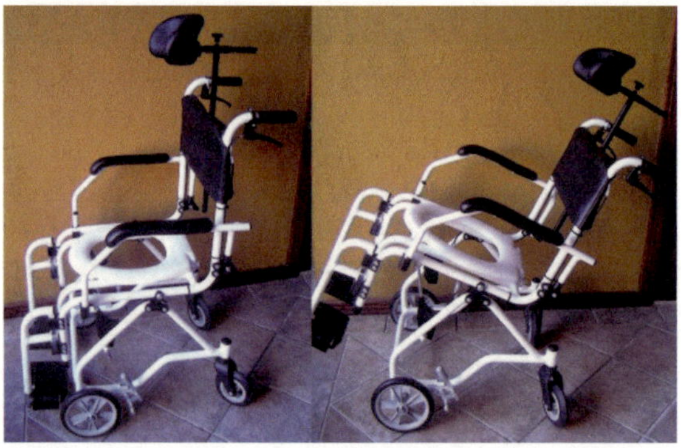

Fig. 4. Aluminum bath chair with backrest and headrest tilt system.

Figure 5 represents another model of bath chair that presents a possibility to "tilt" and lean back, what allows the reclination of the seat/backrest group or just the backrest, a choice that depends on the motor requirements imposed by the disabled person and/or the caregiver. The equipment height might be adjusted via a hydraulic system, promoting an excellent body mechanics to the caregiver. Besides such characteristics, there are other ones, such as adjustable headrest, locking casters, height adjustment of the footrest and armrest, that completes the equipment specifications.

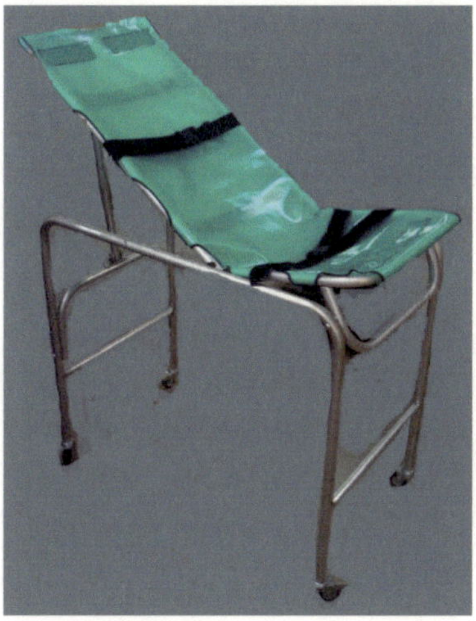

Fig. 5. Bath chair with recline tilt system.

Among these AT devices aimed for bathing or showering that have specific features to the demands imposed by more severely handicapped users, there are other models whose dimensions meet more specifically children and teenagers. These models, in their own constitution, usually have a range of resources appropriate to meet some specification, such as lack of motor control, including absence of neck control, insufficiency or default of trunk control, persistency of primitive reflexes and/or involuntary movements difficult to control, a situation that is compatible to people that with cerebral palsy or other disease that seriously compromise the musculoskeletal system.

Since those people are dependent to caregiver to run bath or shower due to their severe motor disability, some models of chair have adjustable height, whose function is, in addition to providing accommodation to the user through their seat and back features, to provide satisfactory positioning to the caregiver, that is, to the one who performs the task of bathing or showering, and to ensure a safer transference of handicapped people.

The bath chair shown in figure 6 meets the specifications previously described. This type of chair can be used without the base and may be kept on the floor of a bathtub or shower, or with the base, avoiding that the caregiver remains crouching during showering or bathing. In relation to the physical characteristics, this product has a tubular aluminum monobloc structure, epoxy paint, seat and back with tailored double coats (the first part in twisted nylon and reinforced and the second one in a sanlux type canvas, aiming to hygiene, comfort and durability), holes for water drainage during bath or shower, waterproof fabric with adjustable via Velcro seat belts for chest and thigh safety, foam injection headrest covered with vinyl that allows only height adjustment as it is attached to the backrest, four swivel wheels with brakes, in addition to a non-tumble front and rear system. The backrest has ten tilt adjustment controls in the seat made by a spring locking device.

It is worth emphasizing that the presence of casters at the base of the chairs indicated to adult and youth may result in facilitating the movement of the equipment and the user in the shower and other rooms of the houses.

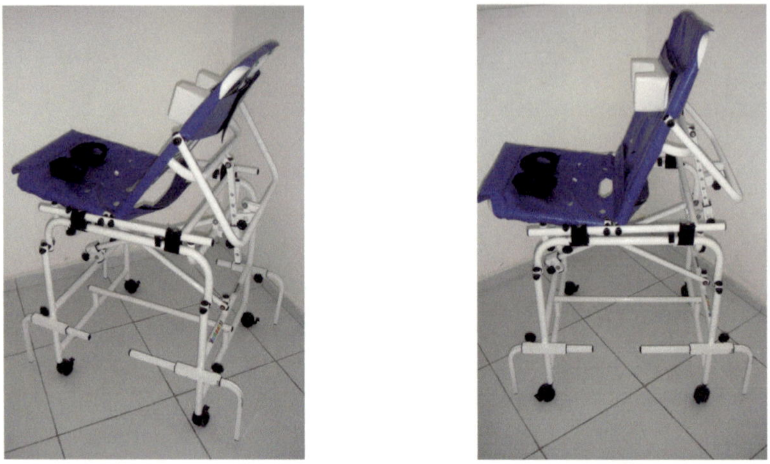

(A) (B)

Fig. 6. Youth bath chair: (A and B) Removable base with height adjustment and belt and headrest system.

Other baht chair models (Figures 7 and 8) are designed to serve severe motor impairment users also have similar aspects, differentiating in some structural features, design and accessories. They are products whose tubular structure is often made of aluminum and the seat and back set consists of a polyurethane shell form-fitting, with or without holes, whose function is let the water flow during bath or shower.

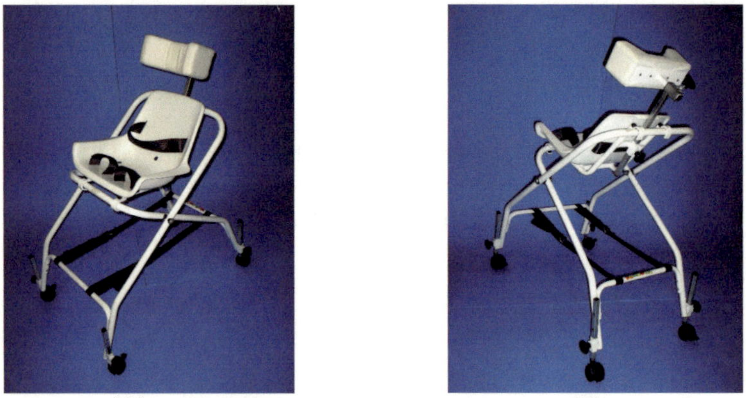

(A) (B)

Fig. 7. Infant bath chair: (A) anterolateral view; (B) posterior-lateral view.

The metal frame of the bath chair in figure 7 (A and B) presents a system of opening and closing in "X", which is caught by two parallel nylon straps with possibility of adjustment, a seat belt made by a chest band and two in the hip whose adjustment and setting are made through a system of Velcro. The presence of the headrest externally attached to the back of the chair allows its height and depth adjustment, which can influence the choice of the equipment due to its contribution for a better user positioning. The headrest (optional) set on the back (Figure 8) only allows height adjustment that is obtained by displacing it onto the back. Moreover, the possibility of height adjustment of seat/backrest in relation to the ground (Figure 7) allows a better adjustment of the equipment to the height of the user that performs the shower of bath task. The present of brake caster makes movement easy and maintain the security of the user.

Added to these products that are part of the AT devices made by large companies in the branch, it is possible to list some equipment in the Brazilian context that present the same purpose as mentioned above, but with its own characteristics, made by artisans or home business companies.

The bath chair presented in Figure 8 is a device produced by a small business, made of painted steel, with a seat and back system made by a plastic sheet and non-slip fabric. It has an "X" locking system concerning its width, while its length remains unchanged after being closed. As it is a tailored-made product, it is possible to get it according to the user measures, caregiver height and dimensions of the environment in use. The presence of foot casters for favoring moving is optional, although, because the seat and backrest are fixed, it is often necessary its transportation to be dried and cleaned.

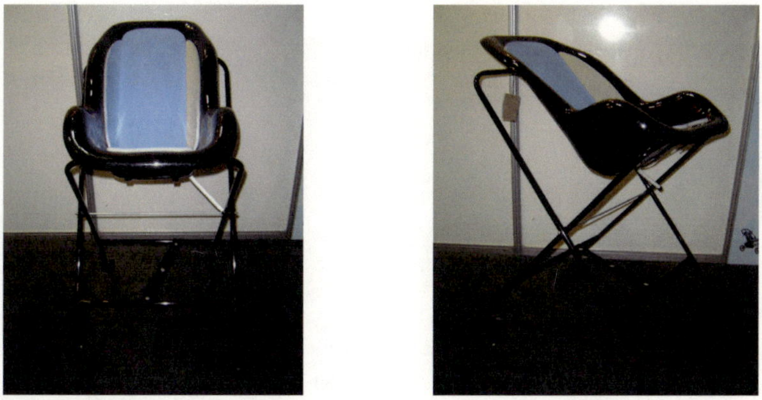

(A) (B)

Fig. 8. Bath chair: (A) anterior view; (B) side view.

Figure 9 presents another device, homemade, whose instructions are available in hospitals and rehabilitation centers, also indicated to attend specific demands of severely handicapped people.

Through explanatory leaflets delivered by those rehabilitation centers to the companions of handicapped people, it is possible to achieve the device in specialty material and fabric stores. Fitting a device up should be done step by step according to the instructions, being the model available in three pattern sizes, that is S, M and B. The structure of the chair consists of connected PVC pipes. For the seat and back it is suggested a non-slip fabric or any washable resistant cloth, such as nylon. It should be noted that the fabric should be attached to the structure (Figure 10).

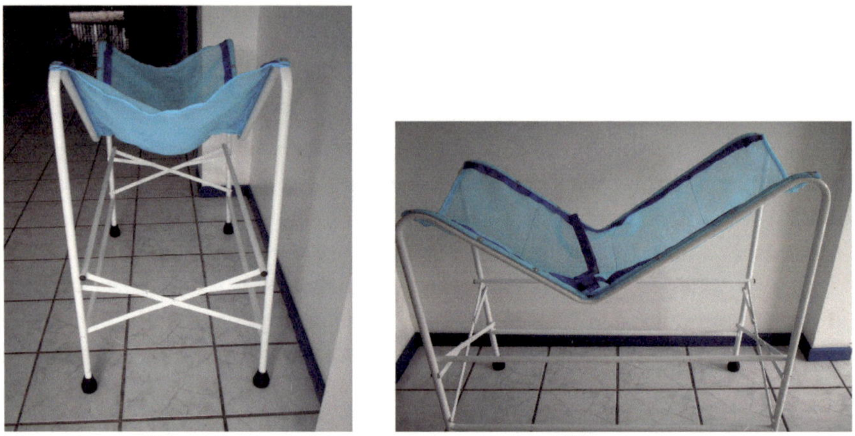

(A) (B)

Fig. 9. Bath chair: (A) anterior view; (B) side view.

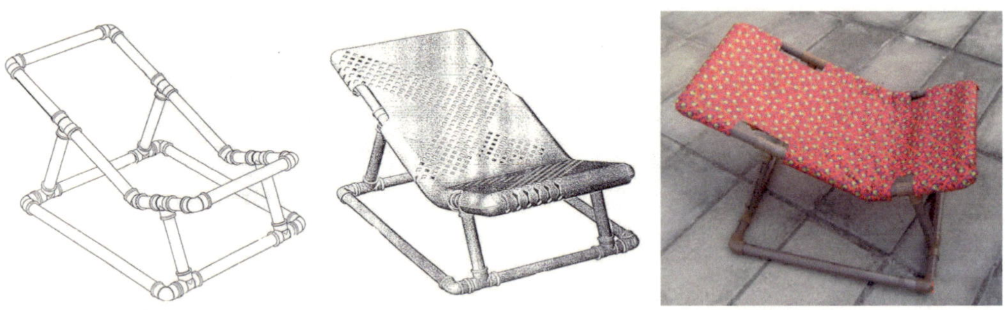

(A) (B) (C)

Fig. 10. (A) and (B): PVC pipes and connections structure; (C) Final craft model.

6. Conclusion

This chapter sought to present some directions for the prescriptions of AT devices, such as bath chairs, available in Brazil, for disabled people, especially physical commitment. Besides, it tried to provide, for the ones who work with handicapped people, relevant information regarding AT equipments that exist to maintain sitting posture during bath or shower.

In today's market there are numerous AT devices designed to meet the requirements demanded by a handicapped person or bath or shower activity. Ideal physical characteristics for usability of an AT product lead to professionals involved in an extensive knowledge of equipments available in the market in order to make it possible a critical analysis in relation to them, providing elements to an adequate indication of the device.

Assistive technology devices used for showering or bathing should accumulate a group of individual characteristics of handicapped people, that is, their structural and functional needs. The set of characteristics analyzed on AT devices should include: (1) product sizes, such as seat width and depth, height of backrest and to the ground; (2) presence of accessories, like head support, belts and other ones; (3) possibility of adjustments, such as height regulation and variation in the angle of the seat/backrest to get a tilt or not-tilt system; and (4) any additional information that enriches analysis for a better indication, prescription and purchase of the equipment.

7. References

American Occupational Therapy Association (2008). Occupational therapy practice framework: domain and process. *American Journal of Occupational Therapy*, Vol. 62, No. 6, (November/December, 2008), pp. (625-683), ISSN 0272-9490.

Baxman Jaguaribe (2011). Produtos - Cadeira de Rodas. *In: Linha Higiene*, 15 abr 2011, Available from: <http://www.baxmannjaguaribe.com.br>.

Comitê de Ajudas Técnicas/CAT. Projeto de Lei do Senado N° 111, de 2007, In: *Presidência da República. Secretaria Especial dos Direitos Humanos. Coordenadoria Nacional para Integração da Pessoa Portadora de Deficiência/CORDE*, 20 may 2011, Available from <http://www.acessobrasil.org.br/CMS08/seo-principal-1.htm>.

Expansão (2011). Produtos. *In: Cadeira de Banho*, 18 may 2011, Available from: < http://www.expansao.com>

Gooch, H. (2003). Assessment of bathing in occupational therapy. *British Journal of Occupational Therapy*, Vol. 66, No. 9, (September, 2003), pp. (402-408), ISSN 0308-0226.

Ortobrás (2011). Produtos - Cadeira de Rodas. *In: Linha Higiene*, 18 may 2011 Available from: < http://www.ortobras.com.br>

Ortopedia Ortorio(2011). Ortorio. *In: Serviços*, 18 may 2011, Available from http://ortorio.com.br/antigo.

Pain, H.; Mclellan, D.L. (2003). The relative importance of factors affecting the choice of bathing devices. *British Journal of Occupational Therapy*, Vol. 66, No. 9, (September, 2003), pp. (396-401), ISSN 0308-0226.

Rede Sarah de Hospitais de Reabilitação (2009). *Apostila de orientação aos pais para construção da cadeira de banho*. Sarah, Brasília.

Santana, C.; Elui, V.; Andrade, V. (2009) Reflections about learning and teaching assistive technology in Brazil. *Technology and Disability*, Vol. 21, No. 1-6, (July, 2010), pp. (143-148), ISSN 1055-4181.

Shepherd, J. (2001). Self-care and adaptations for independent living, In: *Occupational therapy for children*. Case-Smith J., pp. (489-501), Mosby, ISBN 9780323028738, St Louis.

Tabaquim, M.; Lamônica, D. (2004). Análise perceptual de mães e filhos com paralisia cerebral sobre a atividade de banho. *Arquivos Brasileiros de Paralisia Cerebral*, Vol. 1, No. 1, (September/December, 2004), pp. (30-34), ISSN 1807-4456.

Vanzetti (2011). Catálogo Virtual de Produtos. In: *Cadeira de Banho*, 18 may 2011, Available from http://catalogo.vanzetti.com.br.

10

Imaging Systems in Assistive Technology

Miloš Klíma and Stanislav Vítek
Czech Technical University in Prague
Czech Republic

1. Introduction

The working age population, which is conventionally defined as aged between 15 and 64 years, would start to decline as of 2010 and, over the period of 50 years, it would drop by 15% in the EU. In 2060, there would be more than twice as many elderly than children. In 2008, there were about three and a half times as many children as very old people (above 80). In 2060, children would still outnumber very old persons, but by a small margin: the number of very old people would amount to 80% of the number of children.

These numbers are very significant – more money will be spent on care for the sick and elderly people. And what is far more important, there would be not enough labor force, which should take care of these people. In this situation can help the assistive technologies, covering a broad range of different techniques and approaches streaming to the same aim – support and assistance for elderly and handicapped persons. It is a typical interdisciplinary field highly demanding not only towards designers and manufacturers but also towards users.

The labor shortage can be largely solved by applied multimedia imaging systems. Remote monitoring can reduce the amount of recurring admissions to hospital, faciliate more efficient clinical visits with objective results, and may reduce the length of a hospital stay forindividuals who are living at home. Telemonitoring can also be applied on long-term basis to elderly persons to detect gradual deterioration in their health status, which may imply a reduction in their ability to live independently. In this chapter follows general overview and possible applications of such a sytems and particularly inteligent CCTV (Closed Circuit TV) surveillance systems for monitoring of handicaped people, and an aids for everyday use, utilizing multimedia means in any form.

2. Imaging system in general

The imaging system in general is demonstrated in the block diagram on Fig. 1. The first part of system is formed by an optical system. The optical system transforms 3D information from the scene into 2D planar projection on an image sensor (2D distribution of irradiation or illumination). In some cases there are some additional elements inserted between the optical system and sensor – such as image amplifier, shutter, various filters etc. The image sensor performs a conversion of optical projected image into its electrical representation – electro-optic conversion. The image sensor is a 2D array of pixels (picture elements) and supply electrical signal to one or several outputs. The video electrical signal passes through the preprocessing and source coding blocks. Final video signal can be transmitted through the communication channel or retrieved. The separate block is an image processing block providing the selected processing tools – motion detection, object detection and identification, object parameter measurement etc. The final block is an image display converting the electrical representation into the 2D or stereoscopic optical image viewed by an observer.

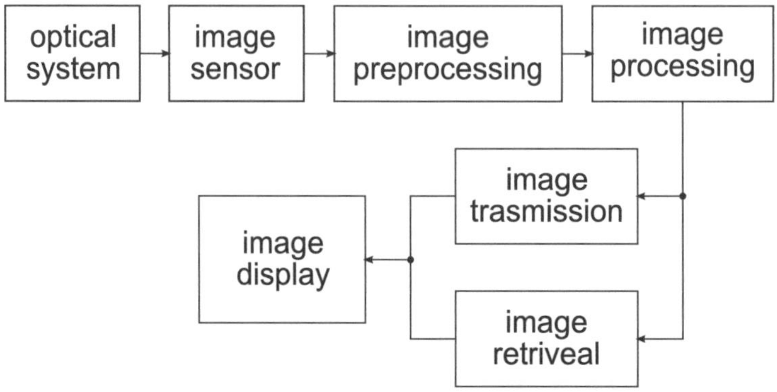

Fig. 1. Imaging system in general

3. Optical system and additional elements

The optical system is formed by an objective – usually as a combination of lenses (refractive optics) but there are some other configurations available – reflective (mirror) optics and diffraction elements. The selection is based upon selected spectral range and particular purpose. Relevant spectral transparency is limiting for lenses, spectral reflexivity for reflective optics. There is a large variety of available materials [X1] for refractive optics e.g. glass for VIS, ZnSe and Ge for IR, quartz and sapphire for UV. In the X-ray range the optical imaging system is constructed as mirror optics. In special terrestrial applications we have to take into account also the transparency of the atmosphere e.g. in X-ray imaging.

The important parameters of optical system are viewing angle, resolution and aberrations. The resolution is frequently described in a number of lines per mm but more precise definition is done by the MTF (Modulation Transfer Function) or PSF (Point Spread Function). The MTF is a module of optical transfer function in spatial domain; PSF is a spatial impulse response of the system. Apart of that the geometrical image distortions (barrel or cushion type) and vignetting are affecting the final image quality. As an integral part of optical system is an aperture. It is characterized by F-number. The function of aperture is multiple. At first it controls the level of sensor illumination (irradiation). The other aperture function is to set the DOF (Depth of Field, Depth of Focus) and the last one affects the resolution (large aperture – larger impact of abberations, small aperture – larger impact of diffraction).

The first additional element is an image amplifier with optical image input and optical image output. There have been used vacuum versions for long time but now the MCP (Microchannel Plate) is widely applied – see Fig. 2.

The other additional elements in the optical system are absorption or interference filters modifying the spectral transparency, polarization filters, etc.

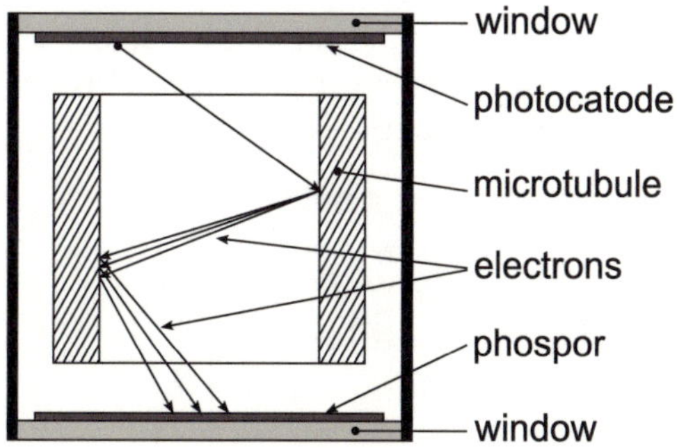

Fig. 2. Microchannel plate.

4. Image sensing

The image sensor is a fundamental part of imaging system. The sensor itself is always a planar array of sensing elements and it means that the sensor performs a 2D spatial sampling of projected image. The 2D sampling can affect the sampled image quality in the same way as it is in 1D sampling in signal domain. The resolution (or MTF and PSF) of optical imaging system should be matched to the spatial sampling raster of sensor otherwise some additional image distortions can be introduced. The spatial sampling frequency value has to fulfill the Shannon sampling theorem (sampling frequency must be at least twice higher than the highest spatial frequency in the image). If not the aliasing effects will appear – very well-known effect called moire. The problem is even more difficult in the color-sensing single-chip image sensors where the color splitting system is created by a CFA (Color Filter Array) in planar version. In this case the coinciding color rasters provides lower spatial sampling frequency than relates to the total pixel number. The example is shown on Fig. 3 – the most frequently used Bayer RGB CFA. Sensor manufactures eliminate this problem by the implementing of 2D anti-aliasing low-pass spatial filter front of the sensor (two optically birefringent plates) in order to limit a maximum spatial frequency in the projected image. For details see Lukac (2009).

Fig. 3. RGB CFA - Bayer structure.

The CFA is realized as an array of thin-film interference filters and their spectral responses are not matched to a particular RGB color system and the RGB signals should be transformed to some commonly used.

Recently there are available two basic types of sensors – CMOS (Complementary Metal-Oxide-Semiconductor) called after the standard memory chip technology and CCD (Charge Transfer Device). The CMOS sensing elements are a PN or PIN photodiodes usually scanned in the XY commuting array structure – see Fig. 4.

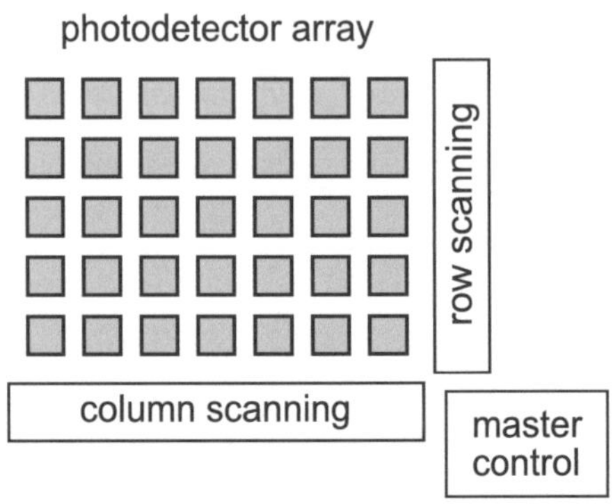

Fig. 4. XY CMOS sensor structure and basic commuting principle.

The CMOS sensors provide high flexibility of scanning sequence (random access, scanning of sensor part) and driving and signal processing circuitry is included in the sensor chip. Moreover the CMOS sensors can offer the function called binning – parallel connection of neighboring pixels in order to increase sensor sensitivity. In order to avoid the problem with the planar CFA 2D sampling the FOVEON® introduced a different vertical color splitting structure where the layers of silicon itself are used as color filters – see Fig. 5.

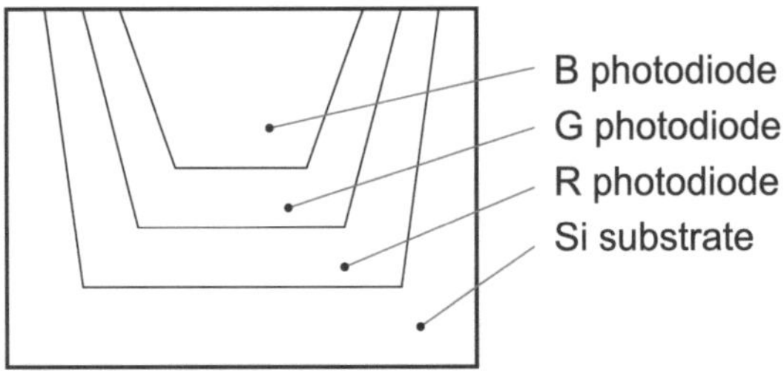

Fig. 5. FOVEON® CFA structure.

CCD sensors are based upon a different principle. The pixel is formed by a MOS capacitor and the signal quantity is a size of charge packet accumulated in this capacitor. The charge packet size is proportional to local pixel illumination. The pixels are organized in rows forming a charge packet accumulating and transferring structure – see Fig. 6. The vertically oriented sensing and transferring lines creates a 2D pixel array. In the FT (Frame Transfer) configuration the 2D charge-packet distribution (frame) is transferred into the similar 2D pixel array. This part is not photosensitive and is used as a memory with joint line shift register array. When the image has been transferred to the memory part the stored image frame is read out line-by-line and pixel-by-pixel. In such a way the image scanning procedure is fixed and cannot be changed and pixels must be read out in given order. Recently the image sensors are manufactured in resolutions from millions to tens of millions pixels.

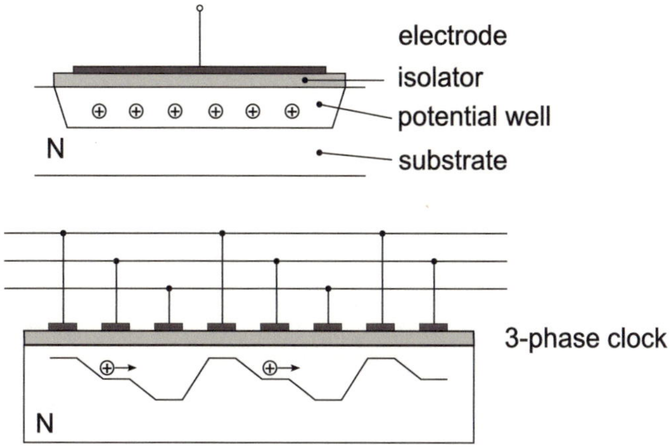

Fig. 6. CCD structure.

Specialized sensors are used esp. in IR spectral regions such as pyroelectric detector array, microbolometric array, microthermocouple array, photoresistor array and also CCD sensors on e.g. HgCdTe semiconductor (different from silicon). Another approach is to convert invisible optical radiation outside of VIS range into VIS applying a proper phosphor or scintillator. It mostly applied for high-energy photons in UV, X-ray a gama spectral range.

5. Image reproduction

Recently image displays are based upon several physical principles. Recent generation displays are LCD (Liquid Crystal Display), plasma, DLP (DMD) and OLED (Organic Light Emitting Diode). The coming generation seems to be FED (Field Emission Display) and SED (Surface Emission Display). Printers dominate the ink-jet and laser types.

The color displays create a color image in similar way as the image sensors split a color image into RGB trichromatic representation. In principle, no trichromatic display with three basic colors can reproduce all existing colors esp. highly saturated. Fig. 7 shows the examples of existing structures. White color is used in some of them to avoid requirement of maximum signals for all three RGB channels and to guarantee correct reproduction of white.

Fig. 7. Structure of reproduction rasters in trichromatic displays.

The LCD pixel functions as an electrically-controlled rotator of light polarization plane. In the absence of external electric field the elongated rod-type molecules of liquid crystal create a twisted structure performing the polarization plane rotation. When the external electric field (driving voltage) is applied the liquid crystal molecules orientation will follow the direction of electric field providing no polarization plane rotation. Combining this element with polarizer(s) provide a required light intensity control – see Fig. 8. The LCD panel can be designed as transparent or reflexive versions.

The LCD does not generate light but can only modulate light from an external light source (it functions as a SLM Spatial Light Modulator or valve). Therefore the LCD uses a luminescent panels or LED array in order to provide required light. LCDs are frequently applied in image projectors where the external light source is usually a xenon-filled discharge tube. The LCD pixel array is multiplexed in a standard XY (column and row) configuration. In TFT (Thin Film Transistor) active displays the LCD pixel array is covered an array of thin film transistors. Each LCD pixel is equipped with one or two TFTs creating a one-bit memory holding information about a required LCD pixel state. It stays until is changed. This technique improves highly LCD performance (brightness, contrast).

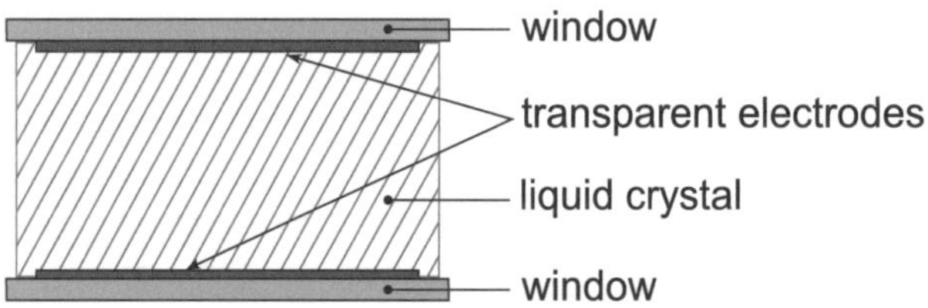

Fig. 8. LCD display principle.

Plasma panel is based upon a pixel array of micro discharge tubes. The discharge tube is filled with a mixture of rare gases e.g. Ne, Xe. The discharge emits UV radiation and excites a layer of phosphor. There are applied different phosphors for R, G and B pixels. Fig. 9 demonstrates the principle. The scanning structure is again XY array but no additional one-bit pixel memory is required. According to the discharge tube VA characteristic there are two stable states – on and off. So the one-bit memory is provided by the discharge itself.

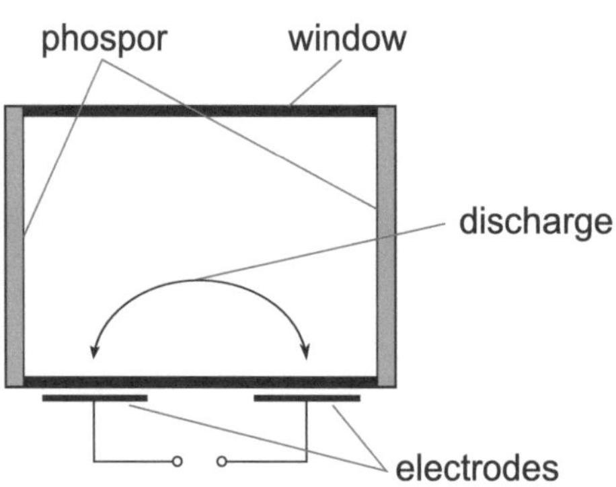

Fig. 9. Plasma display principle.

OLED panels are based upon organic semiconductor materials. In such a material the LED array is created including commuting structure. They are already widely applied in mobile phones and other small screen applications. The advantage of OLED is very wide emitted light spectrum (reproduction of highly saturated colors), transparency of material and mechanical flexibility.

DLP (Digital Light Projector) is equipped with the display chip DMD (Digital Micromirror Device). It is an array of electrostatically deflected micromirrors. Each mirror is controlled separately and its position is stored in a one-bit memory underneath until is changed – for illustration see Fig. 10. The DLP technology is applicable for the image projection to the largest available screens (tens of meters). The DMD is fast sufficiently to enable the time multiplexing of colors instead of isochronal reproduction of colors in all other systems. The time multiplexing of colors was applied in the first-generation color TV cameras before and during World War II.

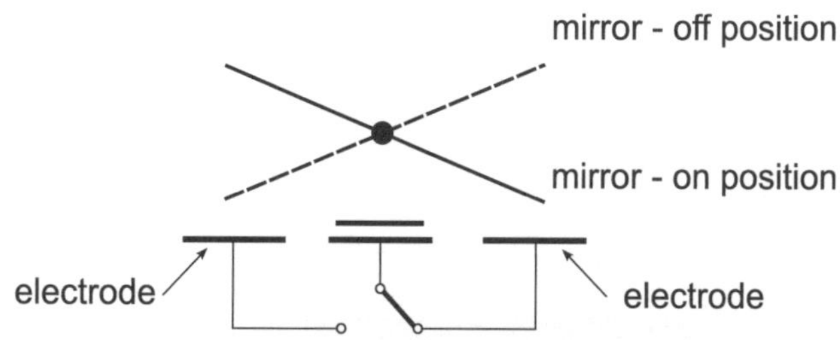

Fig. 10. DLP - DMD principle.

The plasma and DMD have only two states – on and off. Therefore for a gray-scale reproduction the PWM (Pulse Width Modulation) has to be applied for the light intensity control. Moreover using a sequence of impulses of proper length for each pixel one can realize a direct in-pixel D/A conversion.

Near-future displays seem to be FET and SED panels. In both cases it is in fact redesigned principle of old CRT (Cathode Ray Tube) with an electron beam modulation. In all three cases the electron beam is electrically modulated and after that creates an image in a phosphor layer. The source of electrons in CRTs is a heated cathode common for a whole CRT. In FEDs or SEDs there is a cold cathode generating electrons together with a modulation electrode in each pixel. Both principles are demonstrated on Fig. 11 and Fig. 12, respectively.

Stereoscopic (or 3D) image reproduction has been long-term exploited mostly in technical tasks (anaglyph, simulators etc.) and now is entering TV and games fields. The eye-multiplexing can be performed by color, polarization or switched multiplexing goggles/glasses or a display with a raster of cylindrical lenses.

6. Image preprocessing

The image preprocessing includes numerous functionalities to be performed on a raw image. We can list them in the following summary: correction of transfer functions (contrast enhancement) incl. histogram equalization, de-noising, sharpening (edge enhancement), white balance, color system transformation, de-vignetting, etc.

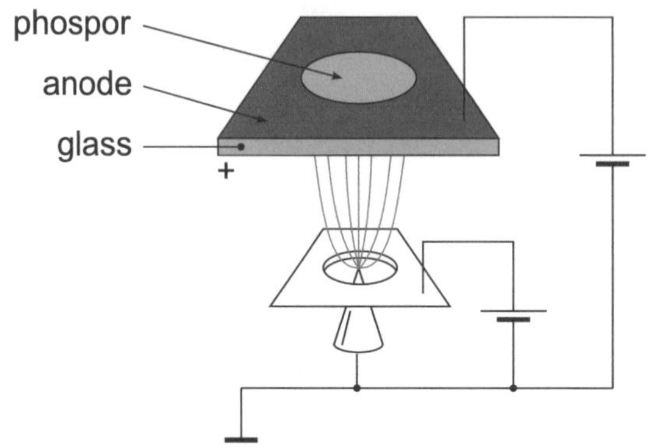

Fig. 11. FED display panel.

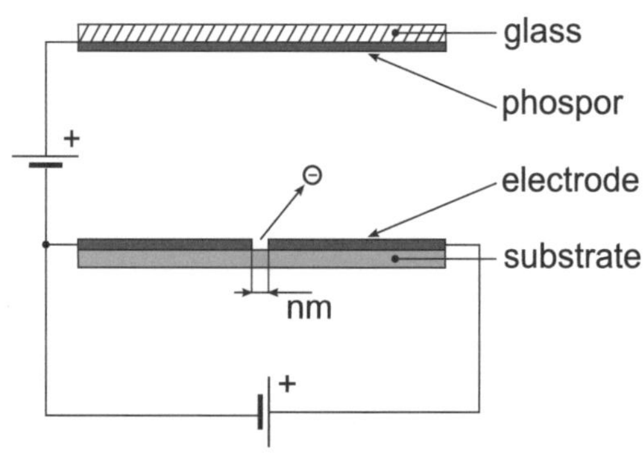

Fig. 12. SED principle.

7. Image/video compression and retriveal

The image compression or source coding is a vital task in imaging systems. The information content (size of file, required bitrate) is very high and it can be reduced significantly. Generally, using lossless compression techniques we can approach a limit by the redundancy. If the compression requirements are higher lossy compression techniques have to be applied. The values of bitrates can be demonstrated on the oldest digital TV standard ITU-R 601 defining SDTV (Standard Definition TV). It requires 216 Mbits/sec for 8bit quantization (720x576 pixels, 4:3, 25 frames/sec, signals Y-13.5 Msamples/sec, R-Y, B-Y - 6.5 Msamples/sec). Recently additional digitizing rasters have been defined - HDTV (High Definition TV, 1k, 16:9), Digital Cinema (2k, 4k) a Super HiVision (8k).

At the beginning of color TV in forties and fifties in the last century several analog source compression standards were developed. Based upon the physiology of HVS (Human Visual System) the basic color signals RGB (required bandwidth 3 x 6 MHz) have been transformed into a combination of the luminance signal Y and two chrominance signals e.g. R-Y, B-Y. The bandwidth requirements of R-Y and B-Y signals are reduced significantly (Y - 6 MHz, R-Y, B-Y - 1.6 MHz). These input signals are consequently coded in a complex color TV signal. Two chrominance signals are modulated by a QAM (Quadrature Modulation) on a color subcarrier and then frequency multiplexed with the luminance signal. The analog coding standards are NTSC (National Television System Committee, USA and Japan) and PAL (Phase Alternating Line, Europe). The SECAM (Sequential a Memoire) originated in France and uses FM modulation of two color subcarriers (France, Eastern Europe). The simplicity of analog source codec schemes leads to continuing exploitation in cable TVs and especially in specialized CCTV systems for technical use but they are disappearing from broadcasting.

Recently, the vast number of digital image compression standards has been developed and is applied. There are still-picture compression standards (JPEG, JPEG2000, TIFF, GIF, PNG, etc.) and video compression standards (MPEG2, MPEG4/Part 2, MPEG4/Part 10/H.264/AVC, WMP, etc.). Most of them are using lossy compression techniques leading to some additional distortions and artifacts. Fig. 13 demonstrate typical example of distortions of JPEG standard.

(a) Original image

(b) Distorted image

(c) Detail of original image

(d) Detail of distorted image

Fig. 13. Example of distortions of JPEG standard.

The image compression standards are in fact procedures for applying image compression tools (techniques, methods). The first compression tool is given by a sampling raster. In color images the full resolution is provided only in the RGB representations. Otherwise the above mentioned signals Y, R-Y, B-Y are used. The full resolution is done in the Y channel only and the R-Y and B-Y signals are sampled in reduced sampling pattern (2 x lower in x direction, 2 x lower in y direction = 4 x in total). Among others there are two most important methods: DCT (Discrete Cosine Transform) with quantization and motion prediction. In all standards an image is split into blocks (JPEG, MPEG1,2 - 8 x 8 pixels, H.264/AVC variable size) at the beginning. The DCT is applied on each block and it results into the set of spectral coefficients. The DCT concentrates energy (lowers redundancy) around the DC component and significantly reduces a number of coefficients to be transmitted. Each coefficient is consequently divided by relevant element of quantization matrix of same dimensions as the original block. The value of quantization matrix elements is perceptually optimized according to the subjective importance of each spectral coefficient. In such a way the number of coefficients is further reduced. Applying a zig-zag reading procedure for coefficients avoids a transmission of the most zero values. Therefore, the block information amount is reduced significantly and is content-dependent. Possible modification of quantization matrix gives a tool for image quality and required bitrate adjustment. In video the motion prediction technique is applied. The initial idea is to describe motion in image as a motion of blocks. The position of particular block is analyzed in the current and following blocks. The block position in another frame is searched through a search window and evaluated by selected block matching criterion. The block motion is than described by a motion vector. There are numerous other image compression techniques applied - VLC (Variable Length Codeword, e.g. Huffman coding), RLC (Run Length Coding), DPCM (Differential Pulse Code Modulation), WT (Wavelet Transform, JPEG 2000) etc.

8. Image/video transmissionn

As already mentioned there are many applications of imaging systems where the analog image compression and transmission is still applied. The analog video transmission is based on same principles as video broadcasting. The complex NTSC resp. PAL signal is VSB-AM (Vestigial Sideband – Amplitude Modulation) modulated on a particular RF carrier forming a TV channel. Because of historical reasons the structure of TV channels is different in different in different regions (USA and Japan, Western Europe, Central and Eastern Europe).

In eighties of the last century the digital TV started and there are two main standards – ATSC (USA) and DVB (Digital Video Broadcasting). They applied the MPEG 2 coding scheme and H.264/AVC coding scheme mostly for HDTV transmission. The ATSC uses the digital 8-state VSB-AM modulation. DVB has developed several broadcasting standards: DVB-T (terrestrial), DVB-S (satellite), DVB-C (cable), DVB-H (handheld), DVB-SH (satellite handheld) etc. Some of these standards are available in second generations: DVB-T2, DVB-S2. These standards differs in a channel coding scheme called FEC (Forward Error Correction) and type of RF carrier modulation according to the properties of particular transmission channel. Two examples have to be mentioned – DVB-T with OFDM (Orthogonal Frequency Division Multiplex) compensating an influence of multipath RF propagation and enabling a SFN (Single Frequency Network) transmitter operation. The other one is DVB-C accepting really multistate modulation up to 256 QAM. Another communication channel applied for image/video transmission is ADSL/VDSL approach trough a twisted-pair metal line. Internet has brought the IP TV streaming.

9. Distortions and artifacts, quality metrics

Image or video information is subjected to a broad variety of distortions during sensing, processing, compression, retrieving and transmission that image quality is a crucial question.

In assistive technologies we mostly expect an image assessment by a human observer but some approaches of machine vision can be implemented too.

There are two basic attitudes to the image quality assessment. The first one is purely oriented to the subjective perception and emotional impact of image - related to subjective quality. Generally, it is defined as QoE (Quality of Experience) against QoS (Quality of Service) usually defined purely technical. The second one can be called as a "feature-preserving" quality – related to detection, identification and classification of objects and events in image/video information. In both cases a lot of effort has been done in modeling of dependencies of subjective quality on some objective image parameters. The most reliable technique for the evaluation of image/video quality is a subjective assessment by a group of observers. There are numerous standards defined by ITU, ETSI, and SMTPE etc. describing practical procedures to be followed in subjective tests. On the other hand, there are numerous objective criteria expressing the image quality – from the simplest ones as MSE or PSNR up to much more sophisticated as SSIM (Structural Similarity Image Metric or VIF (Video Image Fidelity). These metrics can be classified as Full Reference, Reduced Reference and No Reference based upon the availability of original (reference) image.

10. Camera networks

Imaging system itself can only simply monitor the relatively limited space with no possibility to further processing. In order to be truly useful for assistive purposes, it's necessary bring imaging systems together into organized networks and systematically analyze if possible in real time video or image sequence and detect major events in terms of assistive applications.

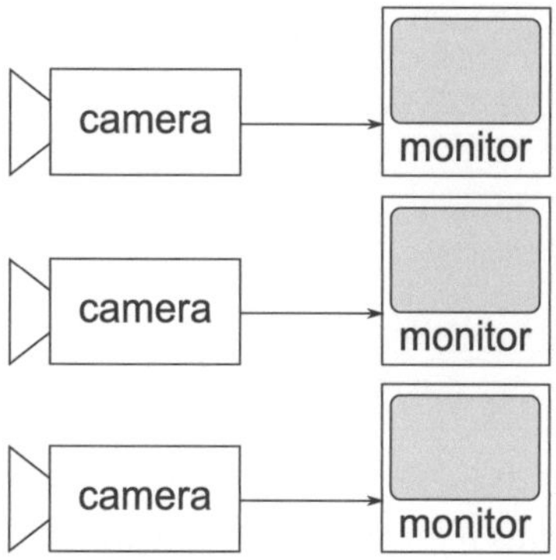

Fig. 14. CCTV system.

The simplest closed-circuit television (CCTV) system shown in Fig. 14 allowed an operator to observe roughly tens of different locations from many viewpoints at once. Systems can use both fixed on moving (PTZ, Pan-Tilt-Zoom) cameras – with PTZ cameras operator can change view of interest if necessary. More complex, centralized system where video from cameras is stored in at a central server which also distributes video for analysis, and to the user through a computer interface is shown in Fig. 15. The most sophisticated decentalized architecture is shown in Fig. 16. So called *smart cameras* are carried out video storage and data processing so bandwidth requrements are reduced and system with distributed metadata can be accessed from multiple locations.

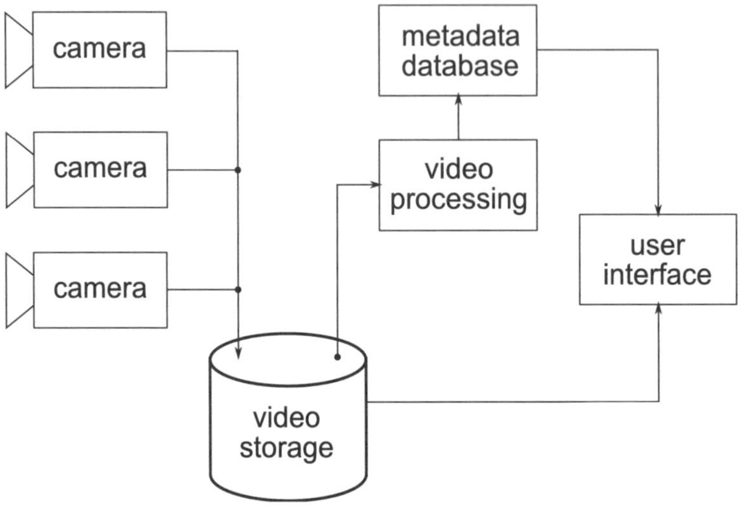

Fig. 15. Centralized architecture.

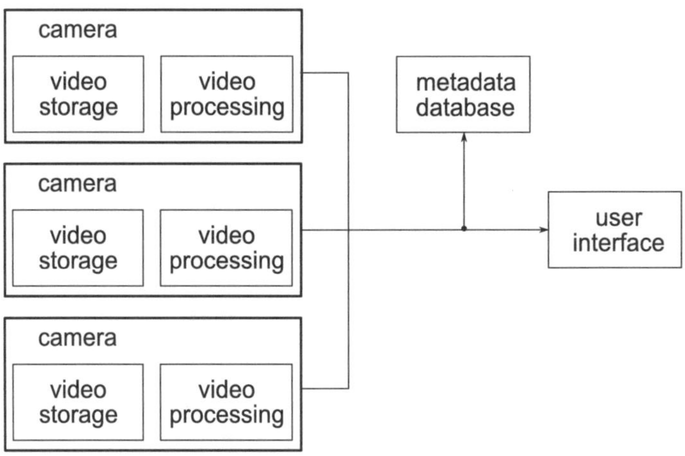

Fig. 16. Decentralized architecture with smart cameras.

A calibrated multicamera networks are useful to provide information on the distance between any object and camera. From this information 3D position of objects can be calculated. Also better models of objects (for example human) can be obtained. However, correct spatial information relies on the automatic matching between corresponding pixels in each image. This process is computationally expensive, for example pixels in low texture areas are very hard to match.

Imaging systems for assistive applications may not have the highest possible resolution. Often is more important to decrease bandwidth, also images with lower resolution are less difficult in terms of image processing. As an example we can mention system CARE (Safe private homes for elderly persons), wireless low-cost stereo vision system using cameras with resolution of 304×256 pixels, which is deployed to use in Germany and Finland Belbachir (2010).

11. Monitoring / video analysis

Monitoring and video analysis for assistive purposes generally means recognition of human activities and generation of action or alarm in case of unexpected behavior of the monitored person. The problem of human action recognition is quite complicated but with adequade choice of image processing methods is possible to find model of articulated non-rigid body. We aim to recognize five types of human daily activities: lying, sitting, standing, walking and other movements including transitions between sitting and standing or lying, and some leg movements when the human subject is sitting or lying – these movements are not assumed to be comparable to walking.

Generally it's possible to split problem of video analysis into following five points:

1. Extract some visual features from images or video sequence. Such features cover wide range of indicators from relatively simple (colour patterns, edges, histogram) to more complex (foreground/background estimation, segmentation, optical flow).
2. Tracking these features in video to obtain temporal sequence using Kalman filters or particle filters.
3. Classifying objects including variations within one class of objects (for example different human poses). For the object recognition can be used for example Haar classificators.
4. Recognition of spatio-temporal patterns like above mentioned lying, sitting, etc. For the tracking of human motion are used points or blobs based models – see Fig. 17.
5. Storing of obtained information as metadata linked to the original image or video data.

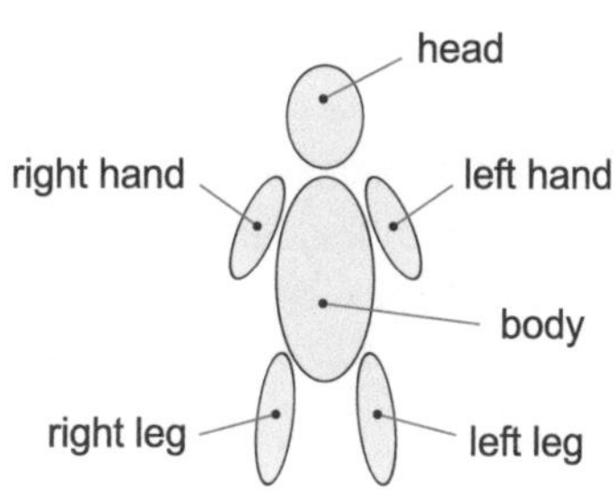

(a) Stick-figure human model.

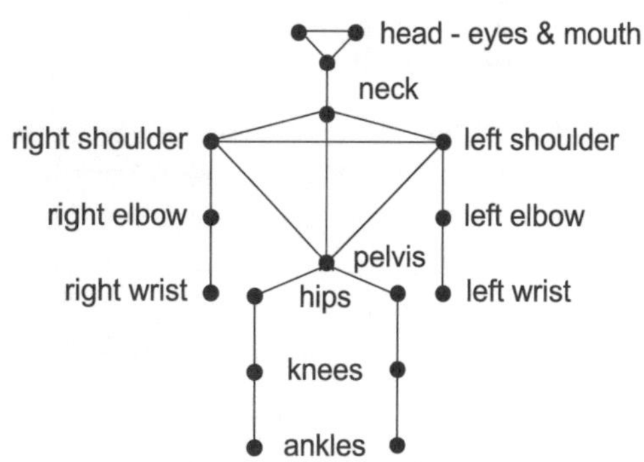

(b) Blob-based human model.

Fig. 17. Models of human body.

For the detection of various kind of objects could be successfully used Viola-Jones object detection framework Viola & Jones (2001), which is able to provide competitive object detection rates in real-time. It can be trained to detect a variety of object classes: human face, hand, upper body etc. During detection phase of the method a window of the target size is moved over the input image, and for each subsection of the image the Haar-like feature is calculated. (Simple Haar-like feature can be defined as the difference of the sum of pixels of areas inside the rectangle, which can be at any position and scale within the original image. Each feature type can indicate the existence or absence of certain characteristics in the image, such as edges or changes in texture.) This difference is then compared to a learned threshold that separated non-object from objects. Beacuse such a Haar-like feature is only weak classifier, in the Viola Jones object detection framework the features are organized in so called *classifier cascade* to form a strong classifier.

For human activity or behavior recognition, most efforts have been concentrated on using state-space approaches. State space models have been widely used to predict, estimate, and detect signals over a large variety of applications. Most common used model is perhaps the Hidden Markov Model (HMM), which is a probabilistic technique for the study of discrete time series. Model defines each static posture as a state. These states are connected by certain probabilities. Any motion sequence as a composition of these static poses is considered a tour going through various states. Joint probabilities are computed through these tours, and the maximum value is selected as the criterion for classification of activities.

Another approach is to use the template matching technique to compare the feature extracted from the given image sequence to the pre-stored patterns during the recognition process. The advantage of using the template matching technique is its inexpensive computational cost; however, it is relatively sensitive to the variance of the movement.

The autonomous monitoring systems should be able to generate alarms. These system can be *rule based* (system is hard programmed) so that when condition or combination of conditions satisfies a certain rule, then apropriate alarm is triggered) or *learning based* – in this case system accumulates typically probability functions to represent what occurs frequently, when an outlier is detected an alarm can be triggered. More informations about automatic video processing can be found in Velastin (2009).

12. Safety & security

Fall is one of the most common causes of injury of seniors. Fractures healing due to falls is bad for old people and therefore is very important to prevent fals (for example by analysis of human balance) or generate an alert if fall happened. Fall-down incident normally occurs suddenly within approximate 0.45 to 0.85 seconds, see for example Chen (2010). When that happens, both the posture and shape of the victim change as shown in Fig. 18. It means that fall detection can be simply based on bounding box aspect ratio measurement.

For the identification of particular persons in the image may be successfully used face recognition based on the three main algorithms: PCA (Principal Component Analysis), LDA (Linear Discriminant Analysis), and EBMG (Elastic Bunch Graph Matching).

With PCA, probe image and training data of the same size must be normalized to line up the eyes and mouth of the subjects within the images. The PCA approach is then used to reduce the dimension of the data and decompose the face structure into orthogonal (uncorrelated) components known as eigenfaces. Each face image may be represented as a weighted sum of the eigenfaces; a probe image is compared against a training data by measuring the distance.

LDA is a statistical approach for classifying samples of unknown classes based on training samples with known classes. This technique aims to maximize between class (i.e. across users) variance and minimize within-class (i.e. within user) variance. When dealing with high dimensional face data, this technique faces the small sample size problem that arises

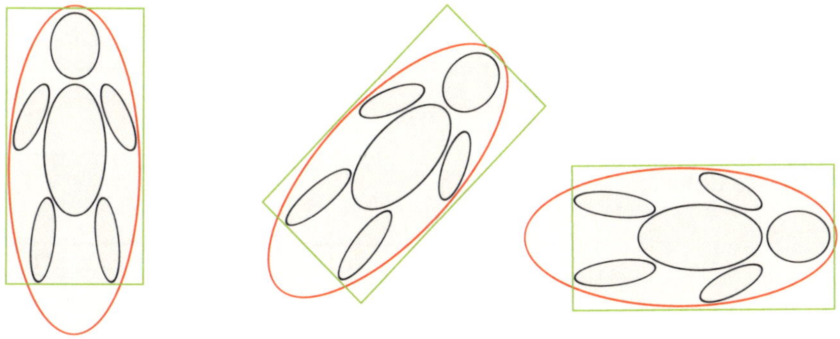

Fig. 18. Phases of fall-down incident.

where there are small number of available training samples compared to the dimensionality of the sample space.

EMGM relies on the concept that real face images have many nonlinear characteristics that are not addressed by the linear analysis methods discussed earlier, such as variations in illumination, pose and expression. A Gabor wavelet transform creates a dynamic link architecture that projects the face onto an elastic grid. Then recognition is based on similarity of the Gabor filter response at circles around nodes of this grid. EMGM method is sometimes combined with PCA or LDA.

13. Combination with other sensors – smart home

The multimedia imaging systems and related technology, which we mentioned in the text, can be effectively utilized in smart homes. Smart home, for example see Fig. 19, is the place where are combined various kind of sensors in order to control people who live here. Outputs of those sensor are processed either automatically or generates alerts for supervising center.

The first of the additional sensors is RFID (Radio-frequency identification). The purpose of RFID is mainly for the fast identification of the persons in the database (i.e. know persons), for example for room access control or identification of the use of potentially hazardous things. Persons have a bracelet that includes a passive most probably unique RFID tag. The RFID readers are typically distributed at the doors of the rooms – when a person crosses a door, the RFID reader detects the bracelet and identifies the person who wears it. Besides to this identification functionality, each RFID reader controls the operation of the electronic lock of the door, allowing or not the access to the room depending on the identity of the person that is going to use it.

Other additional sensor, which works mainly locally as well as RFID, is PIR sensor. This sensor is useful particularly as a detector of presence – it is important in places where is not necessary to identify a specific person or in places where the presence of cameras might be annoying, such as a bathroom, where privacy of the person must be ensured.

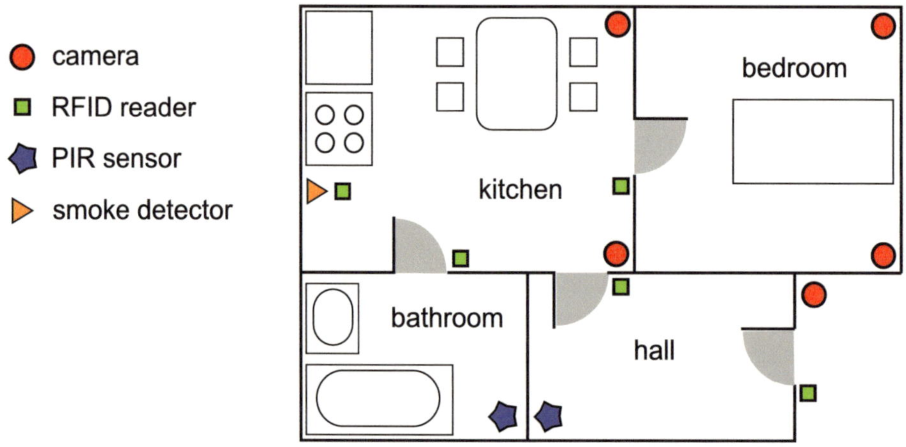

Fig. 19. Example of smart home.

Smoke detector is a typical representative of the group of detectors generating global alerts. Together with the gas detector are used in places where is risk of fire, gas leaks of even explosion.

The last part of smart home we would like to mention is inteligent bed. This bed integrate for example pressure sensors with ECG or other sensors. Simplest usage of pressure sensors in the inteligent bed is monitoring of daily activity of a person: bed-exit time atd. More advanced usage of pressure sensors is measurement of sit-stand movement – it is suitable method for early detection of decreasing mobility, which could help in the prevention of falls. In adition of measurement with pressure sensors is possible to use imaging system also – measured person perform assigned motion, during which is acquired model of human body and the positions of the key points of the skeleton are evaluated – see Fig. 20.

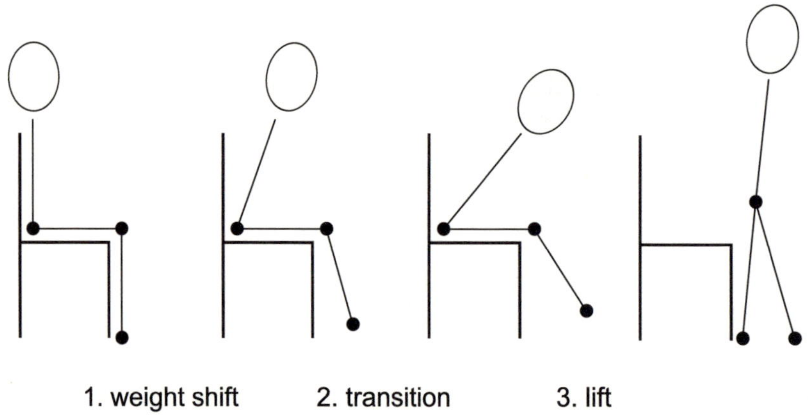

Fig. 20. Phases of sit-stant movement.

14. Conclusions

The Assistive Technology covers a broad range of different techniques and approaches streaming to the same aim – support and assistance for elderly and handicapped persons. It is a typical interdisciplinary field highly demanding not only towards designers and manufacturers but also towards users. In this chapter has been carried out general overwiev of applied multimedia imaging systems. The first part was especially a review of a wide range of available technology for image sensing, reproduction, transmission and compression. Knowledge of the properities of multimedia technology is necessary for the design of assistive application, which is devoted to second part. Processing of video or sequences of images is focused mainly to recognition of human activities, which is together with other sensors base of smart homes – facilities, which will play an increasingly important role in the assistance of our old or handicaped fellow citizens. We can also expect increase of usage of devices like smart phones etc. so data fusion with this kind of devices must be included in design od new assistive technology.

15. References

Lukac, R. (2009). *Single Sensor Imaging*, CRC Press, ISBN 978-1-4200-5452-1.

Fiete, R. D. (2010). *Modeling the Imaging Chain of Digital Cameras*, SPIE Press, ISBN 978-0-8194-8339-3.

Keelan, B. W. (2001). *Handbook of Image Quality*, Marcel Dekker, ISBN 0-8247-0770-2.

Gonzales, R. C. & Woods, R. E. (2008). *Digital Image Processing*, Pearson Prentice Hall, ISBN 978-0-13-168728-8.

Shi, Y. Q. & Sun, H. (2008). *Image and Video Compression for Multimedia Engineering*, CRC Press, ISBN 978-0-8493-7364-0.

Shi, Y. & Real, F. D. (2010). Smart Cameras: Fundamentals and Classification, *Smart Cameras*, Springer, ISBN 978-1-4419-0952-7

Aghajan, H. & Cavallaro, A. (2009). *Multi-Camera Networks, Principles and Applications*, Elsevier, ISBN 978-0-12-374633-7.

Senior, A. (2009). An Introduction to Automatic Video Surveillance, *Protecting Privacy in Video Surveillance*, ISBN 978-1-84882-301-3.

Belbachir, A. N & Lunden, T. & Hanák, P. & Markus, F. & B ottcher, M. & Mannersola, T. (2010). Biologically-inspired stereo vision for elderly safety at home, *Elektrotechnik und Informationstechnik*, Vol. 127, No. 7-8, ISSN 0932-383X.

Aggarwal, J. K. & Cai, Q. (1997). Human motion analysis: a review, *Nonrigid and Articulated Motion Workshop*, IEEE proceedings, pp. 90-102.

Velastin, S. A. (2009). CCTV Video Analytics: Recent Advances and Limitations, *Visual Infomatics: Bridging Research and Pracice*, ISBN 978-3-642-05035-0.

Fan, Q. & Bobbitt, R. & Zhai, Y. & Yanagawa, A. & Pankanti, & S. Hampapur, A. (2009). Recognition of Repetitive Sequential Human Activity, *Computer Vision and Pattern Recognition*, ISBN 978-1-4244-3992-8

Villacorta, J. J. & Val, L. & Jimenez, I. & Izquierdo, A. (2010). Security System Technologies Applied to Ambient Assisted Living, *Knowledge Management, Information Systems, E-learning, and Sustainability Research*, pp. 389-394, ISBN: 978-3-642-16317-3.

Rougier, C. & St-Arnaud, A. & Rousseau, J. & Meunier, J. (2011). Video Surveillance for Fall Detection, *Video Surveillance*, Weiyao Lin (Ed.), ISBN 978-953-307-436-8.

Chen, Y. & Lin, Y. & Fang, W. (2010). A Video-based Human Fall Detection System for Smart Homes, *Journal of the Chinese Institute of Engineers*. Vol. 33, No. 5, pp. 681-690.

Scanaill, C. N. & Carew, S. & Barralon, P. & Noury, N. & Lyons, D. & Lyons, G. M. (2006). A Review of Approaches to Mobility Telemonitoring of the Elderly in Their Living Environment, *Annals of Biomedical Engineering*, Vol. 34, No. 4, pp. 547-563.

Arcelus, A. & Veledar, I. & Goubran, R. & Knoefel, F. & Sveistrup, H. & Bilodeau, M. (2011). Measurements of Sit-to-Stand Timing and Symmetry From Bed Pressure Sensors, *IEEE Transaction on Instrumentation and Measurement*, Vol. 60, No. 5, pp. 1732-1741.

Viola, P. & Jones, M. (2001). Robust Real-time Object Detection, *Second International Workshop on Statistical and Computational Theories of Vision – Modeling, Learning, Computing, and Sampling*.

11

Soft and Noiseless Actuator Technology Using Metal Hydride Alloys to Support Personal Physical Activity

Shuichi Ino[1] and Mitsuru Sato[2]
[1]*AIST – The National Institute of Advanced Industrial Science and Technology*
[2]*Showa University*
[1,2]*Japan*

1. Introduction

In an aging society with a declining birth rate, there is an increased need for home-based rehabilitation systems and human-centered robots for healthcare services. In particular, elderly patients who are required to lie in a reclined position due to stroke or bone fracture may suffer from disuse syndromes, such as bedsores, joint contracture and muscular atrophy (Bortz, 1984). It is difficult for these elderly patients to actively exercise for preventive rehabilitation. Thus, to manage these disuse syndromes, rehabilitation equipment and assistive devices, such as bedside apparatuses for the continuous exercise of joints and power assistance devices for standing and transfer hoists, must be developed. This equipment demands powerful and soft actuators in addition to human muscles (Bicchi & Tonietti, 2004).

However, there are currently no commercially available actuators with the desired characteristics, that is, human compatibility, softness (for safety), noiselessness and a high power-to-weight ratio. These technical requirements present a challenge to the development of rehabilitation and personal autonomy systems. To solve this challenge, various types of artificial muscle-like actuators, such as pneumatic muscle actuators, shape memory alloy actuators and polymer actuators, have been actively studied by material scientists and biomedical engineers. However, satisfactory artificial muscle technologies for rehabilitation or healthcare devices have not yet been realized.

In order to fulfill the above demands for the design of a human-friendly actuator, we originally developed several actuator systems using metal hydride (MH) materials as a flexible mechanical power source. MH actuators can, even in small and light packages, produce a powerful and soft force, because MH materials can store a large amount of hydrogen gas by controlling heat energy; this energy is about a thousand times larger than the volume of the MH alloy itself. Moreover, MH actuators have human-compatible traits, such as softness and noiseless motion, which are derived from the reversible thermo-chemical reaction of metal hydrides. An additional potential merit is that hydrogen is a clean energy carrier candidate because it does not adversely affect the environment (Sakintuna et. al., 2007).

The purpose of this chapter is threefold: first, to overview the properties of MH materials for MH actuators; second, to describe the structure of MH actuators and a performance upgrade that installed a soft bellows made of a laminate film; and third, to show several applications of MH actuators in rehabilitation and assistive technologies that require a continuous passive motion (CPM) machine for joint rehabilitation and a power assistance system for people with restricted mobility.

Finally, we propose future work related to the further improvement of MH actuators to obtain more suitable devices in rehabilitation engineering and assistive technology.

2. Metal hydride alloys

MH alloys have the particular ability of storing a large amount of hydrogen, which can be approximately 1,000 times as large as the volume of the alloy itself. One of the conventional MH materials is a magnesium-nickel alloy (Mg_2Ni), which was discovered in 1968 at the Brookhaven National Laboratory in the USA (Wiswall & Reilily, 1974) and was the first practical MH alloy. Shortly thereafter, in 1970, a lanthanum-nickel alloy ($LaNi_5$) was discovered at the Philips Research Laboratories in the Netherlands (Van Mal et al., 1974). These successive discoveries triggered research on various MH alloys and opened new possibilities for their industrial development.

Moving from historical to scientific background information, metal hydrides (MH_x) are created by a chemical reaction between an alloyed metal (M) and hydrogen gas (H_2), as in the following thermochemical equation:

$$\mathrm{M} + \frac{x}{2}\mathrm{H}_2 \leftrightarrow \mathrm{MH}_x + Q \qquad (1)$$

where Q is the heat of formation, and thus, $Q > 0$ J/mol for H_2. If this reaction proceeds at a fixed temperature, then a hydrogen pressure will advance up to an equilibrium pressure, which is called the plateau pressure. The PCT diagram (P: hydrogen pressure, C: hydrogen content and T: temperature) shows the basic characteristics of an MH alloy. As demonstrated in the PCT diagram in Fig. 1, controlling the temperature of an MH alloy can successfully change its plateau pressure. An MH alloy that absorbs hydrogen well for actuator applications would have a flat and wide plateau area in the PCT diagram. Moreover, the hydrogen equilibrium pressure is given as a function of the temperature of an MH alloy by the van't Hoff equation. Consequently, the logarithmic equilibrium pressure is proportional to the inverse function of the temperature, as shown in Fig. 2.

Furthermore, MH alloys are not combustible, so they are safe as hydrogen storage materials for fuel cells in road vehicles and other various mobile applications (Schlapbach & Züttel, 2001). Thus, we have determined that MH alloys should be used as the power source of actuation for assistive technology.

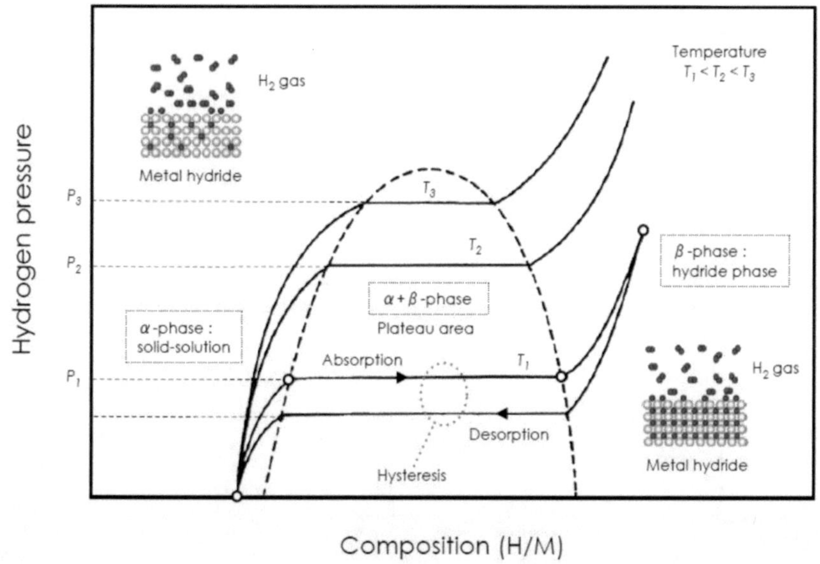

Fig. 1. Pressure-composition-temperature (PCT) diagram of a metal hydride (MH) alloy under hydrogenation.

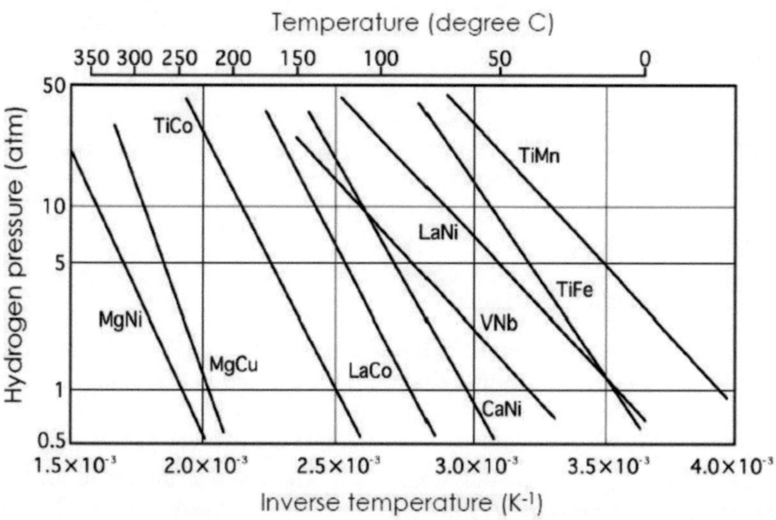

Fig. 2. Relationship between the equilibrium hydrogen pressure and the temperature in various MH alloys.

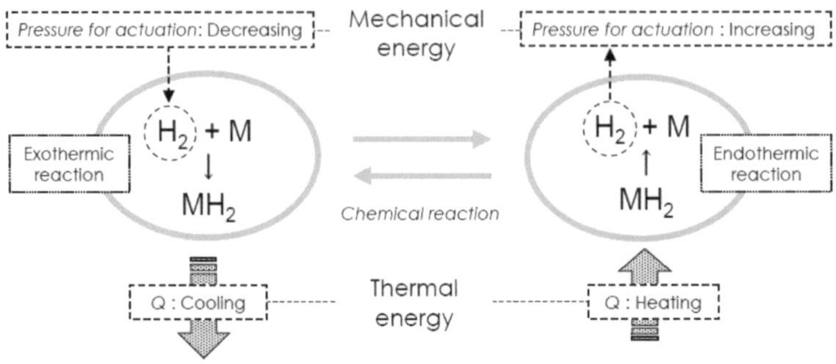

Fig. 3. Schematic illustration of actuation principles based on an energy conversion mechanism between heat and hydrogen pressure in an MH actuator.

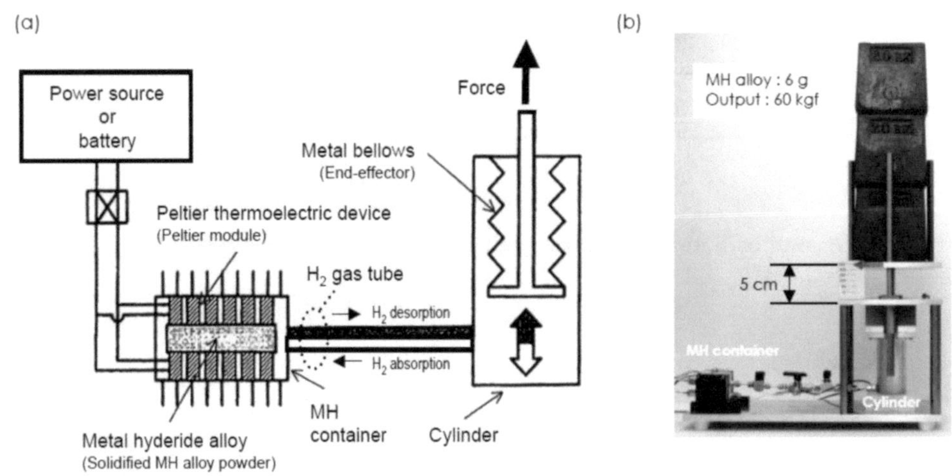

Fig. 4. (a) Basic structure of an MH actuator and (b) the assembled MH actuator using a metal bellows and a Peltier thermoelectric device.

3. Metal hydride actuators

3.1 Basic structure and characteristics

MH alloy actuators can not only store a large amount of hydrogen gas efficiently, but they also desorb hydrogen gas by controlling the temperature of the MH alloy. If this reversible chemical reaction is carried out in a hermetically closed container system, the heat energy applied to the MH alloy is converted into mechanical energy via an equilibrium pressure change in the container system, as shown in Fig. 3. The basic structure of an MH actuator system is shown in Fig. 4 (a). An MH actuator performs by using the hydrogen pressure generated by the MH alloy through heat energy, which is controlled by a device such as an electrically heated wire, a Peltier module (a thermoelectric heating/cooling device) or a solar collector.

MH actuators are composed of a solidified MH alloy powder, Peltier modules to electrically control the temperature of the MH alloy, a temperature sensor, a container for these elements, and an end effecter to transfer the hydrogen pressure into a driving force (Sasaki et al., 1986; Ino et al., 1992).

For example, the MH actuator system shown in Fig. 4 (b) has a 6-g MH alloy and a 36-mm-diameter metal bellows. This actuator can smoothly lift a 60-kg weight without any noise. The power-to-weight ratio of the MH actuator is also very high compared to those of conventional actuators such as electric motors and pneumatic actuators. However, a few weaknesses of MH actuators are their low energy efficiency and slow speed when Peltier modules are used as the thermal controller. However, the heat drive mechanism of the MH actuator does not produce any noise or vibration. In addition, the reversible hydrogen absorption and desorption of the MH alloy also has a buffering effect to cushion a human body and prevent mechanical overload. Therefore, MH actuators are suitable for use as a human-sized flexible actuator that can be applied to soft and noiseless rehabilitation systems and assistive devices.

3.2 The treatment of MH alloys

To improve the moving speed of MH actuators, MH alloy powders are chemical plated with copper, which conducts heat well, and are pressed during a solidification treatment. The MH alloy powders are coated with a 1.0-μm-thick layer of Cu (20 wt%) and solidified into a plate-like compact MH alloy by pressing them to a 3.0-mm thickness. After both treatments, a temperature control device, such as a Peltier, module is directly attached to the compact MH alloy.

Therefore, the heat conductivity of a Cu-coated MH alloy is approximately 50 times that of a bare MH alloy (Wakisaka et al., 1997). It has been found that the addition of copper yields a clear increase in the hydrogen absorption speed.

3.3 The composition of MH alloys

The $LaNi_5$ alloy is more suitable for use in an MH actuator than other compositions of MH alloys because the region of the plateau pressure in its PCT diagram is adequately flat and broad. Moreover, the hysteresis effect that is shown in the PCT diagram, which is inconvenient for controlling an MH actuator, is relatively small for the $LaNi_5$ alloy.

When MH actuators are used in a living environment, the response speed of an MH actuator depends on the initial setting value of the plateau pressure of the MH alloy at room temperature. Thus, an adjustment of the plateau pressure from the PCT diagram is important to achieve the desired mechanical behavior for each application.

The relationship between the hydrogen equilibrium pressure and the temperature is adjustable by changing the alloy composition with the addition of cobalt and manganese. We have selected two different $LaNi_5$ alloys for a comparative discussion. One is $LaNi_{4.45}Co_{0.5}Mn_{0.05}$ (MH-P), which is designed to have a plateau pressure at room temperature (20 °C) of 0.1 MPa. The other is $LaNi_{4.3}Co_{0.5}Mn_{0.2}$ (MH-N), which is designed to have a plateau pressure at room temperature of 0.05 MPa lower than atmospheric pressure. The PCT diagrams of these MH alloys are shown in Fig. 5.

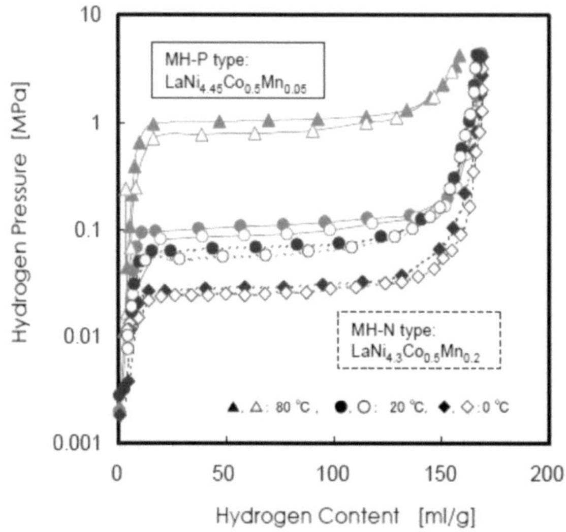

Fig. 5. PCT diagrams of the MH-P alloy (solid line) and the MH-N alloy (dotted line).

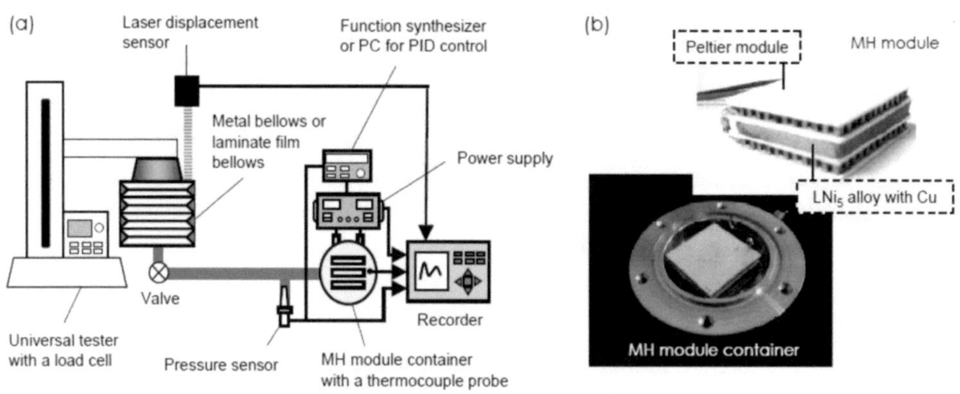

Fig. 6. (a) Experimental setup for the performance testing of the MH actuator and (b) photograph of the MH module and its container.

The experimental setup to measure the mechanical behavior of the $LaNi_5$ alloys (MH-P and MH-N) for the MH actuator is shown in Fig. 6. The $LaNi_5$ alloys were heated or cooled by the Peltier modules that were connected to an electric power supply.

The up-down motion of the MH actuator that was installed with the MH-P alloy or the MH-N alloy is shown in Fig. 7. When the weight of the load was 5 kg, the response speed of the MH-P alloy in the heating phase (+5 V) was approximately twice as fast as that of the MH-N alloy. In contrast, the response speed of the MH-P alloy in a cooling phase (-5 V) was approximately twice as slow as that of the MH-P alloy. When the weight of the load was 0 kg, the response speed of the MH-P alloy in a heating phase was similar to that of the MH-N alloy. However, the response speed of the MH-P alloy in a cooling phase was approximately five times slower than that of the MH-N alloy. It has been found that a plateau pressure from the PCT diagram that is lower than atmospheric pressure can improve the slow shrinking (downward) speed of the cooling phase of an MH actuator (Ino et al., 2009a).

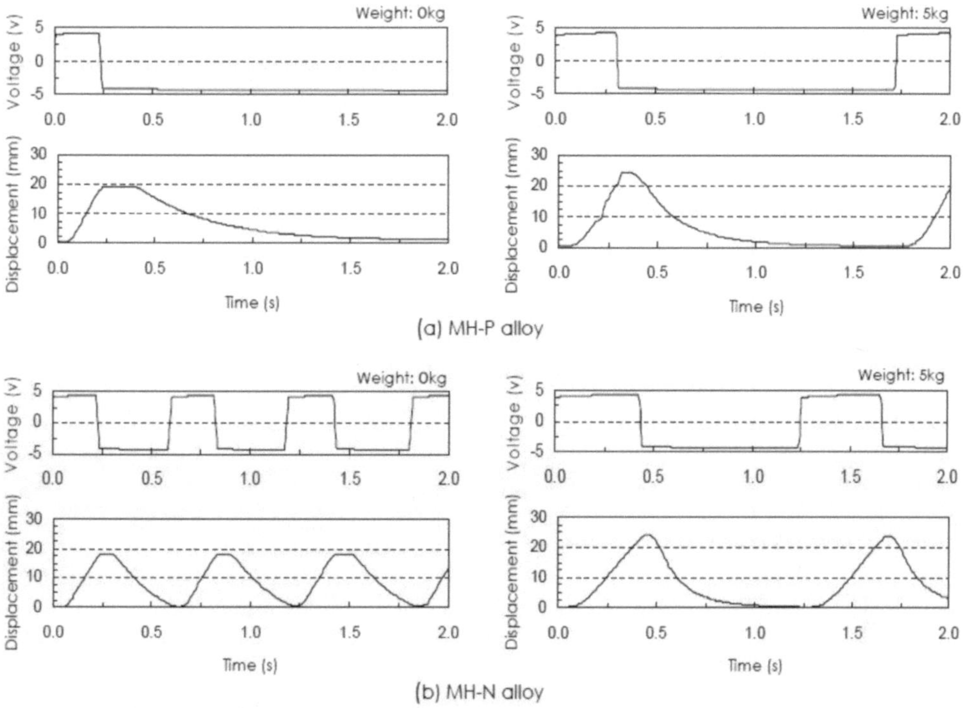

Fig. 7. Time traces of the displacement of the metal bellows supplying a rectangular input voltage to the Peltier elements in the MH module.

3.4 Laminate film bellows

To maintain impermeability against the hydrogen gas from the MH alloy, a metal bellows has been used as the end-effector of some conventional MH actuators. However, a metal bellows is made of stainless steel and is not very suitable, in terms of weight, softness and elongation rate, for rehabilitation equipment and assistive devices. For safety and comfortable usage, rehabilitation equipment and assistive devices must be light and soft compared to the human body. Therefore, instead of the metal bellows, we developed a soft and light bellows element made of a trilayer laminate film of polyethylene-aluminum-polyester (Ino et al., 2009b).

The weight of this laminate film bellows was about 20 times lighter than that of the metal bellows, and the elongation range of the laminate film bellows was approximately 30 times as large as that of the metal bellows. These mechanical properties of the laminate film are very useful in designing a soft MH actuator.

It is well known that a polymer-metal laminate film is a strong gas barrier to oxygen, water vapor and other substances in the packaging industry (Schrenk & Alfrey Jr., 1969). However, no experimental data exist about the hydrogen impermeability of laminate films and their adhesion area by thermo-compression bonding. Thus, the hydrogen impermeability of the laminate film bellows was tested by temporally monitoring the inner pressure and the displacement of a prototyped laminate film bellows that was filled with pure hydrogen gas at 20 °C. The amount of inner pressure and the displacement of the laminate film bellows were maintained over 250 hours, and it was found that the laminate film bellows is capable of maintaining a hydrogen gas barrier for at least ten days.

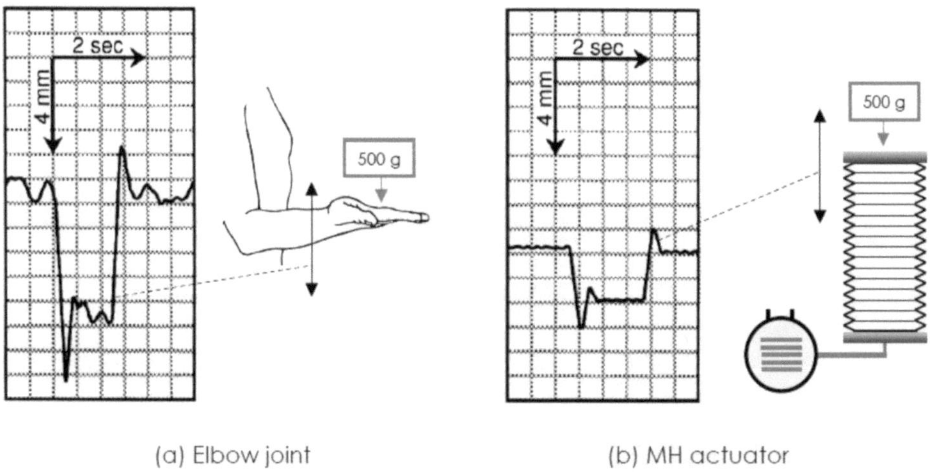

Fig. 8. Dynamic response of (a) the human elbow joint and (b) the metal bellows of the MH actuator under step loading.

3.5 The elasticity of MH actuators

The dynamic elasticity of an MH actuator with a metal bellows and the elbow joint of a healthy subject were compared through the step response by dropping a 500-g weight. Both displacement patterns of the step response were very similar to each other, as shown in Fig. 8. Thus, this compliant feature of MH actuators offers a solution for physical therapy and assistive technology applications.

The elasticity of an MH actuator adopting the laminate film bellows was also measured by a universal tester. The relationship between the stiffness and the strain of the laminate film bellows when the parameters were the initial inner parameters of the MH actuator is shown in Fig. 9. The stiffness increased with increasing strain on the laminate film bellows, and the rate of the stiffness change decreased with the increasing initial inner pressure of the soft MH actuator.

Moreover, the range of the variable stiffness of soft MH actuators with laminate film bellows is smaller than that of human muscle at full activation (approximately 100-200 N/mm; Cook & McDonagh, 1996). Thus, MH actuators are suitable for wearable force display systems, such as rehabilitation equipment and power assist devices, from the viewpoint of mechanical impedance matching and safety in passive elasticity to reduce any potential danger to the human body.

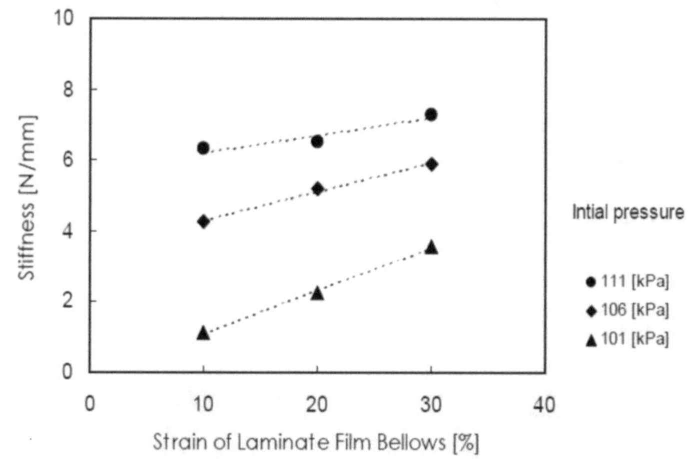

Fig. 9. Stiffness profiles of the MH actuator with the laminate film bellows.

4. CPM machine for joint rehabilitation

4.1 Background

Home rehabilitation devices for elderly people and stroke or joint injured patients are required in an aging society. Some healthcare techniques for joint rehabilitation include manual therapy and range of motion (ROM) exercises using a continuous passive motion (CPM) machine (Salter et al., 1984). The therapeutic effects of CPM, which are the prevention of joint contracture and muscular atrophy, and the increase of ROM after joint injury, fracture and immobilization, have been clinically shown in previous studies (Michlovitz et al., 2004). However, current CPM machines have some problems, such as a lack of the flexibility inherent to the human body, a bulky size for home use and the noise that is emitted from an installed electric motor and gears. These problems preclude ease and safety of use of the CPM machine at home and at the bedside. Hence, we have designed a compact MH actuator and prototyped a CPM device using it.

4.2 The design of a hand and an elbow CPM machine

The prototyped hand CPM machine for a finger joint is shown in Fig. 10. This hand CPM machine was installed with a compact push-pull (double-acting) MH actuator that includes a small metal bellows covered with a stainless steel sleeve and a pair of MH containers with a 3-g MH alloy and a Peltier module. The output force and stroke of the compact push-pull MH actuator were approximately 100 N and 20 mm at maximum, respectively. The weight of the MH actuator was approximately 250 g. The hand CPM machine using this MH actuator was much lighter than conventional CPM machines, which are built up with an electric motor and gears.

To evaluate the dynamic behavior of the hand CPM machine, the MH actuator was given a voltage input to the Peltier module, and the responses were measured. For example, Fig. 11 shows the displacement pattern of the bellows cylinder by a rectangular voltage input applied to the MH module. It was observed that the motion of the push-pull MH actuator was noiseless and smooth within the allowable behavior for joint rehabilitation that was required by a delicate ROM exercise.

We also designed an elbow CPM machine using a pair of metal or laminate film bellows as shown in Fig. 12. In particular, the antagonistic mechanism composed of the double MH actuators using the laminate film bellows allows for compliant action of the human elbow joint and cutaneous surface. This antagonistic mechanism also takes advantage of the feature that its stiffness and position can easily be controlled based on the sum and difference of the inner pressure of both laminate film bellows, respectively (Ino & Sato, 2011).

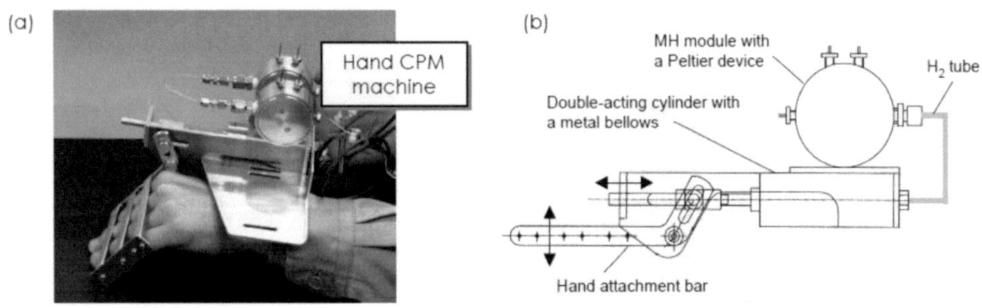

Fig. 10. (a) Photograph of a prototype hand CPM machine for a finger joint and (b) structural drawing of its double-acting MH actuator.

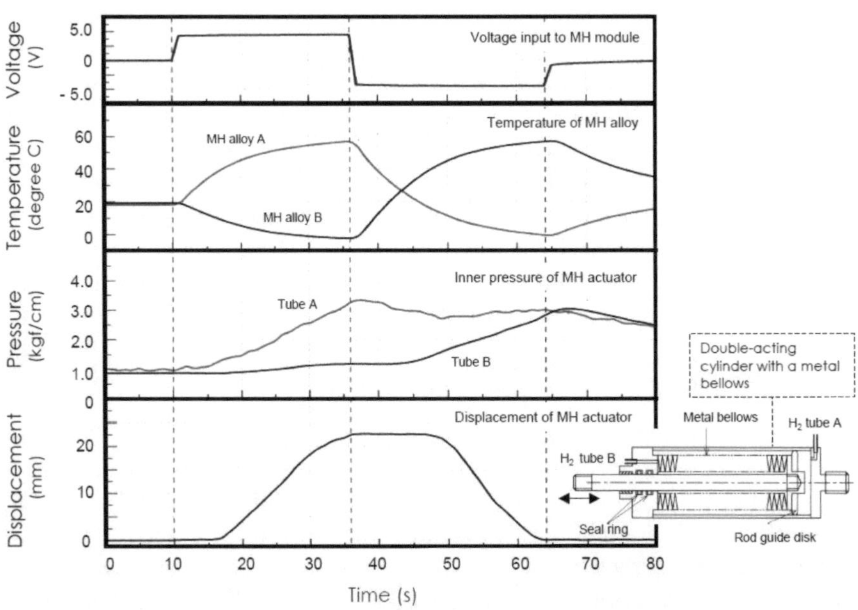

Fig. 11. Dynamic behavior of the compact double-acting MH actuator installed in the hand CPM machine.

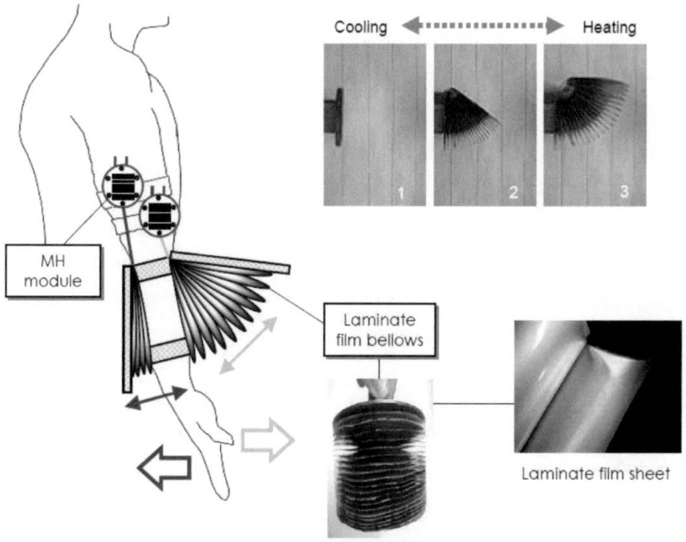

Fig. 12. Design image of an elbow CPM machine using a pair of laminate film bellows and MH modules.

5. Toe exercise system for pressure ulcer prevention

5.1 Background

The pathogenic mechanism behind pressure ulcers observed with bedsores is assumed to be related to defective blood circulation within the subcutaneous tissue (Jan et al., 2006). It is likely that the facilitation of blood circulation in sites that commonly develop pressure ulcers can prevent bedsores. Particularly, the passive exercise of toes may improve the blood circulation in convex portions of the lower limb, such as the lateral and medial malleolus and the first metatarsal head, and heal common pressure ulcers. If a passive exercise system can be applied to the toes and feet, then it is likely possible to reduce the volume of the system and provide a low physical burden to elderly and disabled people who require long-term bed rest.

To assess the usefulness of the passive exercise of toes for the purpose of preventing disuse syndromes, we performed measurements of the subcutaneous blood flow at a common site of pressure ulcers during passive toe exercises. We then developed a prototype motion-assist system for toe exercises using a fan-shaped MH actuator to easily fit various foot forms (Ino et al., 2010).

5.2 The effect of passive toe exercise

The subjects' toe joints were alternately bent and stretched within a specified ROM using a thermoplastic attachment connected to the shaft of a DC servomotor, as shown in Fig. 13. This toe exercise motion induced by the servomotors lasted two minutes with a 6-s cycle time and was initiated after keeping the subjects in a supine position at rest for 10 minutes. Subcutaneous blood flow was measured at the base of first toe and at the lateral malleolus using a laser blood flow meter. The blood flow rate data were acquired for one minute before exercise (pre-exercise), two minutes during exercise (mid-exercise), and one minute after exercise (post-exercise) in each trial. Ten healthy subjects (20 to 80 years of age) participated in this experiment after giving informed consent.

From the experimental results, the blood flow rate at the base of the first toe during the exercise was significantly higher ($P < 0.02$, Friedman test) than before the exercise, as shown in Fig. 14 (a). Similarly, the blood flow rate at the lateral malleolus exhibited the same trend ($P < 0.02$, Friedman test), as shown in Fig. 14 (b).

The subcutaneous blood flow at common pressure-ulcer sites on the foot was increased in spite of a lack of active exercise (self-exercise). One of the factors contributing to increased blood flow during passive toe exercise is assumed to be the mechanical deformation of capillaries in the subcutaneous tissue, due to the extension and flexion of the toe joints. In addition, the effects of the mechanical muscle pump may be due to the cyclical, passive motion of the toe joints.

These results suggest that the continuous passive exercise of toe joints using some type of actuation device may be a valuable healthcare apparatus to help prevent disuse syndromes such as pressure ulcers and soft tissue contracture induced by ischemia.

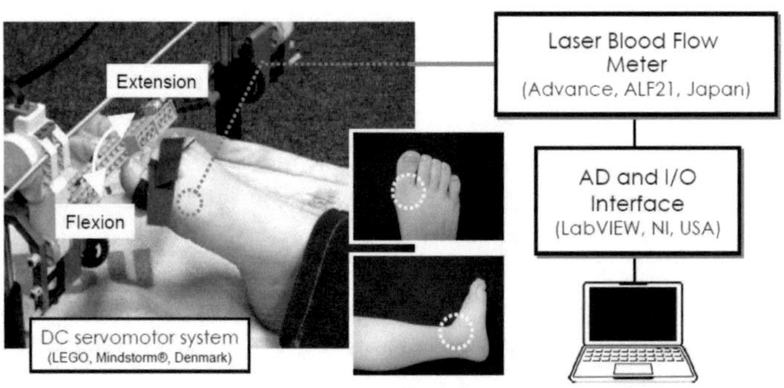

Fig. 13. Experimental setup for the continuous passive exercise of toe joints using a DC servomotor and the measuring site of subcutaneous blood flow in the foot.

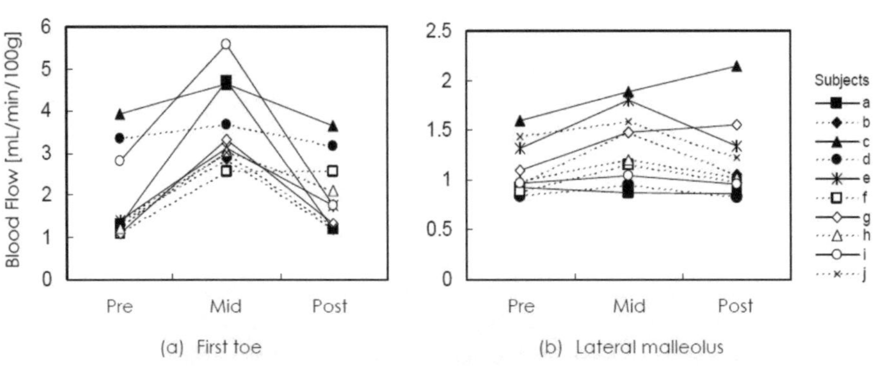

Fig. 14. Subcutaneous blood flow rate of pre-, mid-, and post-passive exercise at the following: (a) the base of the first toe and (b) the lateral malleolus.

5.3 Range of motion and force

The target force and angular ROM required to design the passive toe exercise apparatus were measured in a clinical physical therapy practice, as shown in Fig. 15.

The results of both the maximum applied force and the ROM of the metaphalangeal joint of the toes at flexion and extension during the manual passive exercise by a skilled physical therapist are shown in Fig. 16. The mean maximum force at extension was approximately 35% higher than that at flexion. Meanwhile, the mean maximum ROM at extension was approximately 20% lower than that at flexion. The measured values of the maximum force (flexion: 7-14 N, extension: 9-20 N) and the maximum ROM (flexion: 24-47 deg., extension: 16-40 deg.) also varied with each subject.

It is suggested that passive exercise systems must be adjustable with regards to the output force and the working angle to account for the physical variations of individual patients.

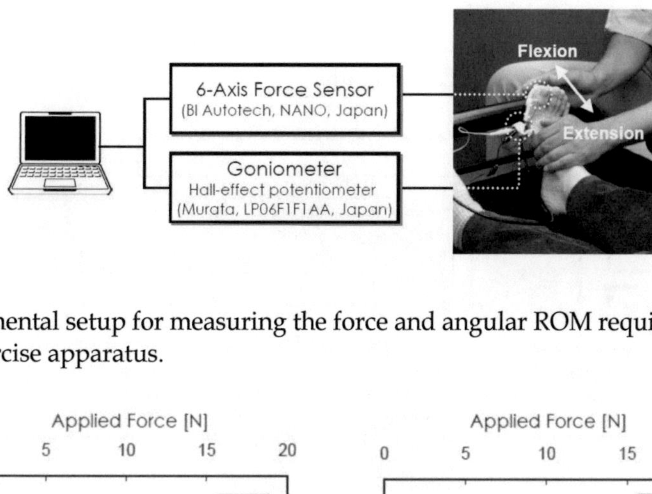

Fig. 15. Experimental setup for measuring the force and angular ROM required to design the passive toe exercise apparatus.

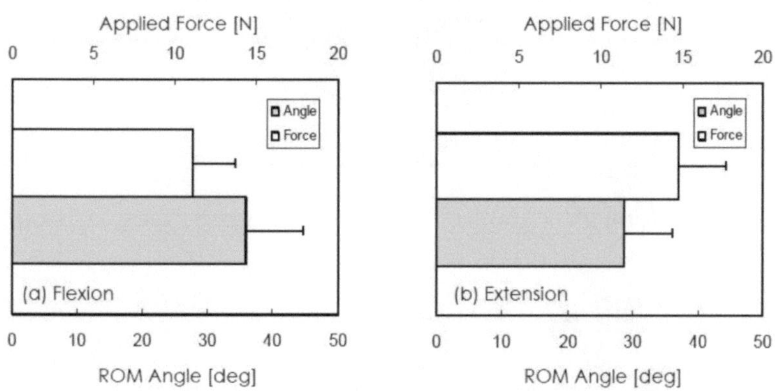

Fig. 16. Maximum force and angle applied manually to toe joints for passive ROM exercise by a physical therapist.

5.4 The design of a toe exercise system

A prototype of the passive exercise system for foot and toe pressure ulcer prevention consisted of two fan-shaped MH actuators with a thermoelectric control device and laminate film bellows, several pressure sensors, a bipolar power supply and a personal computer with a PID controller. A photograph and a block diagram of the prototype apparatus are shown in Fig. 17.

The extension and flexion motion of the toe joints was derived from a pair of laminate film bellows that spread out in a sector form in a plastic case. The motion of the toes in the passive exercise system was gentle and slow enough to facilitate subcutaneous blood circulation. During the operation of the apparatus, the various subjects' toes, of variable size and shape, always fit into the space between the antagonistic laminate film bellows. The mechanical features of the actuator and the toe exercise system are shown in Table 1.

The motion patterns for the joint exercises were produced by the regulation of the pressure combination of both MH actuators. Linear control over the flexion and the extension angle of the toes was obtained by altering the difference between the inner pressures of each laminate film bellows or a pair of the MH actuators. The stiffness of the movable part inserted between both of the laminate film bellows was determined by evaluating the sum of the inner pressures of both bellows. Thus, both the angle and the stiffness of the movable apparatus were easily controlled independently.

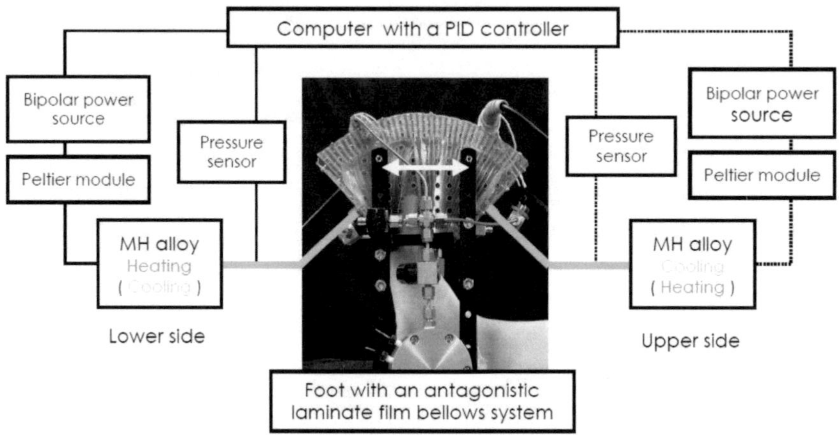

Fig. 17. Block diagram and photograph of the prototyped passive toe exercise apparatus.

	MH actuator with laminate film bellows	Passive toe exercise apparatus
Size	75 mm dia, 23 mm thick / MH container 100 mm dia, 10 mm thick / Laminate film bellows	120 mm x 210 mm x 120 mm
Weight	300 g / MH container 40 g / Laminate film bellows	2 kg
Force (Pressure)	< 100 N (0.03 MPa gauge)	< 30 N (0.1 MPa gauge)
Stiffness	< 10 N/mm	< 10 N/mm
Working range	0-200 mm	± 45 deg

Table 1. Mechanical features of the soft MH actuator using the laminate film bellows and the passive toe exercise apparatus.

6. Transfer assistance system for lower limb disability

6.1 Background

Physically vulnerable people who find it difficult to stand up by themselves due to some illness, injury or age, need help moving between a wheelchair, a bed, a toilet seat, a bathtub or a car seat. Transfer assistance systems for these people are required to be sufficiently safe and powerful to lift their body. In the assisted transfer motion, a physical and mental load exists between the vulnerable people and their helpers at any time. However, securing mobility through transfer assistance tools facilitates personal autonomy in the daily lives of disabled people. Thus, we developed a transfer hoist, which was installed with a high-powered MH actuator, based on an ergonomic motion analysis of transfer behaviors (Tsuruga et al., 2001).

6.2 The analysis of transfer motion

The transfer motion of subjects was simultaneously measured by a three-dimensional motion capture system including three 60-Hz infrared cameras, a force plate, an electromyogram (EMG) telemeter and a personal computer. The subjects were five young men between 20 and 30 years of age and three elderly men of approximately 60 years of age, and they participated in this experiment after giving informed consent.

The muscle activity of the subject's lower limb and the center of pressure (COP) amplitude while standing from a 50-cm height seat are shown in Fig. 18. The standard height of a wheelchair seat is 50 cm in Japan. As a result, the activity of the anterior tibial muscle in the leg was great at 60 degrees and plateaued from 70 to 90 degrees as a function of the ankle joint angle. Meanwhile, the activity of the rectus femoris muscle in the thigh was high at 90 degrees of the ankle angle and plateaued from 60 to 80 degrees of the one. The peak lateral amplitude of the COP indicated that the instability of the standing motion was significantly increased at an ankle joint angle of greater than 70 degrees. These plotted patterns of the muscle activity of the lower limb and the COP amplitude in the young subjects were similar to the elderly subjects. From these results, the suitable angle of the ankle joint at the early standing phase ranged from 60 to 70 degrees. Because joint stiffness is more typical in the elderly than the young, the setting angle of the ankle joint applying a transfer assistance system should be approximately 70 degrees.

We measured the tilting angle of the subject's trunk at seat-off when the angle of the ankle joint was set at 70 degrees. From the experimental results, the tilting angle of the trunk of the young and the elderly was approximately 35 degrees and 45 degrees, respectively.

6.3 The design of a mobile floor hoist

A prototype of the mobile floor hoist for transfer assistance is shown in Fig. 19. The mobile floor hoist had a height of 120 cm and a width of 70 cm. To prevent a fall with the knee flexed during standing and to transfer the user's body stably and safely, the mobile floor hoist had various protective parts, such as a kneepad, an arm pad and a chest pad, which provided a cushioning function. The design of these cushion pads was based on the measurement of the ergonomic motion analysis described above. For example, the kneepad had a variable positioning mechanism to follow the change of the ankle joint angle during transfer motion. Each mechanical part of the hoist adopted a module fashion to adjust easily according to the various physical features of users. A double-acting MH actuator with two MH modules for the lifting mechanism was installed in our mobile floor hoist. The double-acting MH actuator could control its own stiffness and position by changing the balance between the inner and outer hydrogen pressures of a metal bellows in a cylinder. This double-acting MH actuator could lift up to a 200-kg weight to a 35-cm height.

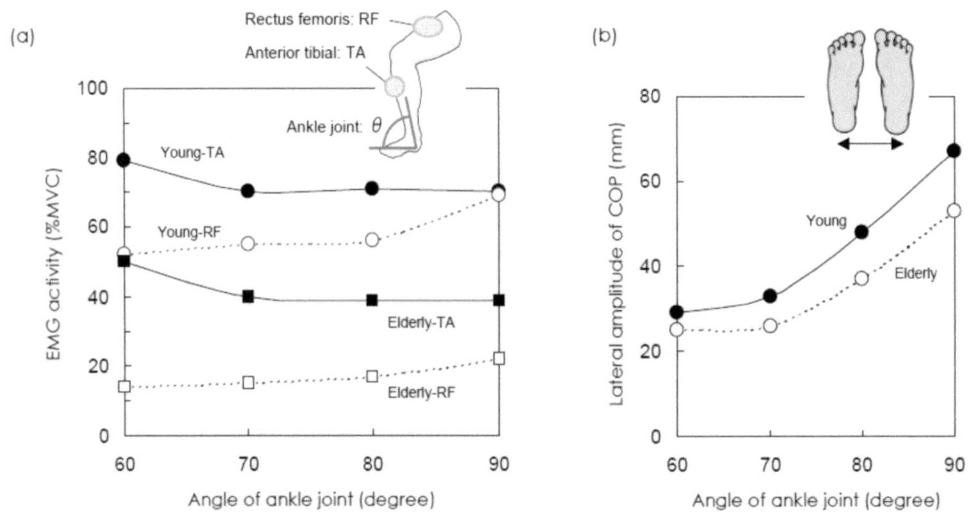

Fig. 18. Comparison of the muscle activity of a lower limb and the COP amplitude while standing from a 50-cm height seat between young and elderly subjects.

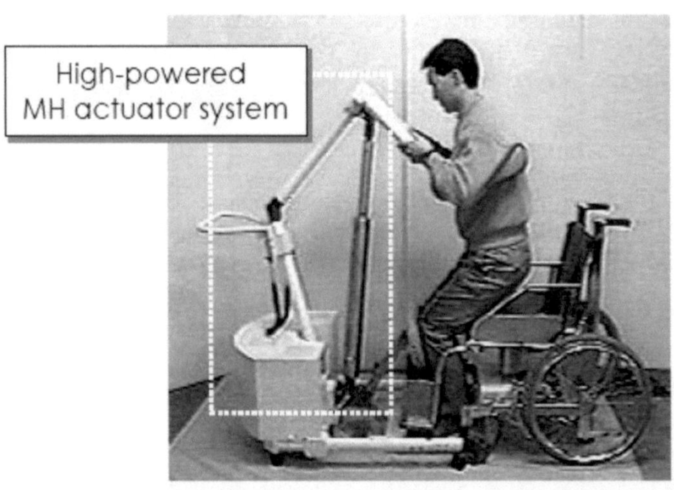

Fig. 19. Photograph of a prototype mobile floor hoist for transfer assistance.

6.4 The design of a seat lifter for a wheelchair or toilet

A wheelchair is one of the most popular pieces of assistive technology equipment for the elderly of people with a lower limb disability. Recently, the wheelchair has been significantly improved, and many types have been developed to conform to the lifestyle of wheelchair users. When using a wheelchair, however, some fundamental tasks that require a standing position, such as reaching a tall shelf and cooking in the kitchen, are still difficult in daily life. Thus, we developed a wheelchair with a seat lifter using an MH actuator to improve the quality of life for wheelchair users.

The assistive devices for lifting the body must secure long and smooth stroke motion for users. Therefore, a hybrid MH actuator that uses a tandem piston cylinder with a hydraulic converter was especially developed, as shown in Fig. 20. When hydrogen gas flows into the lower piston, silicone oil in the lower piston moves into the upper piston, depending on the inflow of the hydrogen gas. By using this hybrid driving mechanism with a hydrogen-hydraulic system, the stroke displacement becomes doubled compared with a typical MH actuator that is driven by only hydrogen gas.

This hybrid MH actuator containing a 40-g compact MH alloy could lift an 80-kg weight to a 40-cm height, as shown in Fig. 20 (b). The lift speed was approximately 20 mm/s, and the total weight of the lifting unit including the MH actuator with the tandem piston cylinder was approximately 5 kg.

A prototype of the toilet seat lifter using an MH actuator is shown in Fig. 20 (a). The toilet with a seat lifting function has a similar hybrid MH actuator system to that described above. This toilet MH actuator system containing a 150-g MH alloy compact could lift a 100-kg weight to a 35-cm height. The lift speed was approximately 10 mm/s. The heater and the cooler of the MH alloy adopted the sheath heater made of electric heating wire and the water jacket utilized tap water in-house to save energy.

This toilet seat lifter was installed in a smart house, the Welfare Techno House (WTH) in Japan, which was built to investigate the daily well-being of elderly people (Tamura, 2006).

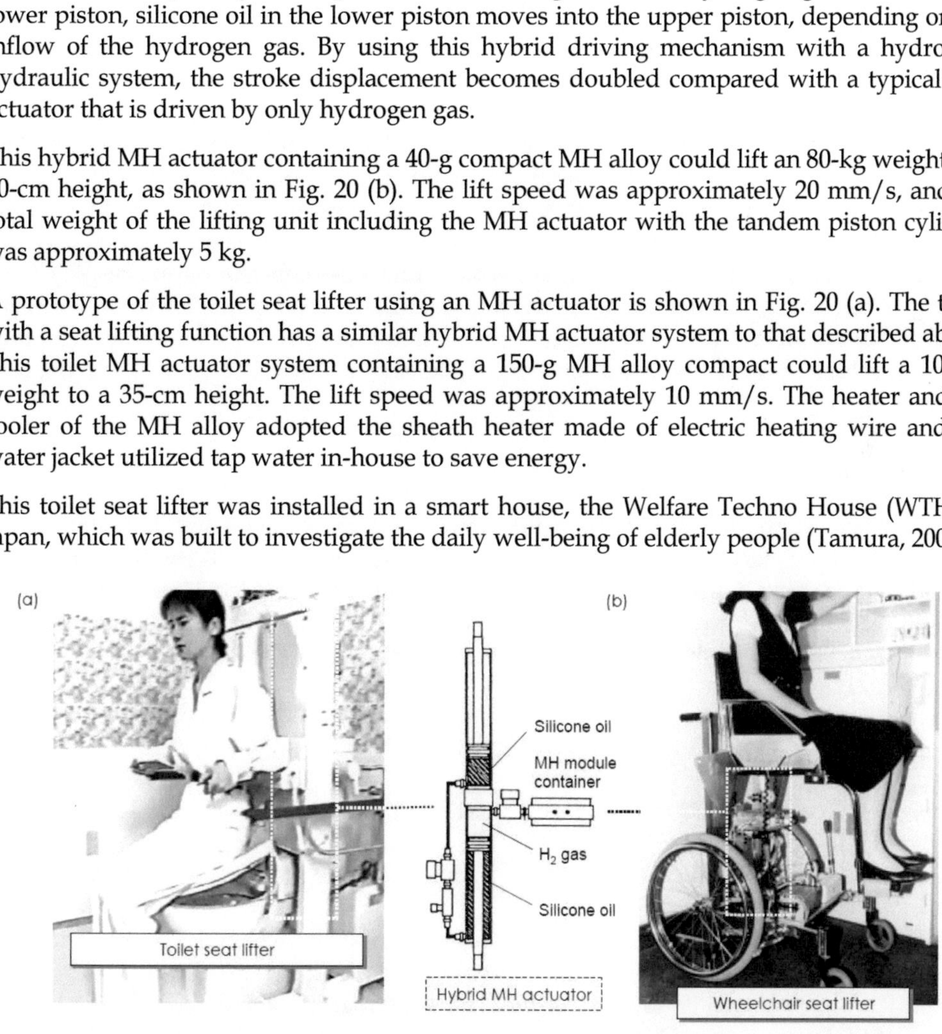

Fig. 20. Photograph of a portable lifter using the MH actuator for (a) a toilet seat and (b) a wheelchair seat.

7. Conclusion

In this paper, we described a soft and noiseless actuator using a MH alloy and its applications in assistive technology and rehabilitation engineering. The MH actuator mentioned above have various unique features, such as a high force-to-weight ratio, low mechanical impedance, noiseless motion and a muscle-like actuation mechanism based on expansion and contraction, which differs from conventional industrial actuators. From these distinctive features, we think that MH actuators are suitable force devices for applications in human motion assistance and rehabilitation exercise.

However, technical challenges also remain regarding the motion speed and energy efficiency of MH actuators, most of which can be overcome through the improvement of Peltier thermoelectric devices (the heat-pump module) and a heat sink material to facilitate waste heat management (Sato et. al., 2011).

Our aim is to develop a new human-friendly actuation device with targeted applications for quality-of-life technology for active health promotion and to facilitate the personal autonomy of elderly and disabled people (Cooper, 2008). While soft and noiseless MH actuators are under development for a CPM machine for joint rehabilitation, a toe exercise system for pressure ulcer prevention and a transfer assistance system for lower limb disability, they can also be adopted for other applications: an active prosthesis and orthosis, a compact wheelchair lift for a low floor vehicle ramp assembly, a haptic display for a sensory substitute or virtual reality. MH actuators have great potential to enhance the physical affinity of assistive technology, rehabilitation engineering and human-machine interface technology, and the sophistication of the MH actuator is a part of our ongoing studies.

8. Acknowledgment

This work was supported in part by a Research Grant from the Okawa Foundation, the Grant-in-Aid for Scientific Research from JSPS and the Industrial Technology Research Grant Program from NEDO of Japan. We would like to thank for Hideto Ito, Hiroshi Kawano, Masahiko Muro, Yuichi Wakisaka, Takeshi Tsuruga, Minako Hosono, Takeshi Hayashi and Shinich Yoshimura for outstanding technical assistance and many fruitful discussions.

9. References

Bicchi, A. & Tonietti, G. (2004). Fast and "soft-arm" tactics. *IEEE Robotics & Automation Magazine,* Vol. 11, No. 2, pp. 22-33

Bortz, W. M. (1984). The disuse syndrome. *Western Journal of Medicine,* Vol. 141, No. 5, pp. 691-694

Cook, C. S. & McDonagh, M. J. N. (1996). Measurement of muscle and tendon stiffness in man. *European Journal of Applied Physiology,* Vol. 72, No. 42, pp. 380-382

Cooper, R. A. (2008). Quality-of-Life Technology; A Human-Centered and Holistic Design. *IEEE Engineering in Medicine and Biology Magazine,* Vol. 27, No. 2, pp. 10-11

Ino, S.; Izumi, T.; Takahashi, M. & Ifukube, T. (1992). Design of an actuator for tele-existence display of position and force to human hand and elbow. *Journal of Robotics and Mechatronics,* Vol. 4, No. 1, pp. 43-48

Ino, S.; Sato, M.; Hosono, M.; Nakajima, S.; Yamashita, K.; Tanaka, T & Izumi, T. (2008). Prototype Design of a Wearable Metal Hydride Actuator Using a Soft Bellows for Motor Rehabilitation, *Proceedings of the 30th Annual International Conference of the IEEE Engineering in Medicine and Biology Society*, pp. 3451-3454, ISBN: 978-1-4244-1815-2, Vancouver (Canada), August 2008

Ino, S.; Sato, M.; Hosono, M.; Nakajima, S.; Yamashita, K. & Izumi, T. (2009a). A Soft Metal Hydride Actuator Using $LaNi_5$ Alloy and a Laminate Film Bellows, *Proceedings of the 2009 IEEE International Conference on Industrial Technology*, pp. 1215-1220, ISBN: 978-1-4244-3506-7, Gippsland (Australia), February 2009

Ino, S.; Sato, M.; Hosono, M. & Izumi, T. (2009b). Development of a Soft Metal Hydride Actuator Using a Laminate Bellows for Rehabilitation Systems. *Sensors and Actuators: B. Chemical,* Vol. B-136, No. 1, pp. 86-91

Ino, S.; Sato, M.; Hosono, M.; Nakajima, S.; Yamashita, K. & Izumi, T. (2010). Preliminary design of a simple passive toe exercise apparatus with a flexible metal hydride actuator for pressure ulcer prevention, *Proceedings of the 32th Annual International Conference of the IEEE Engineering in Medicine and Biology Society*, pp. 479-482, ISBN: 978-1-4244-4123-52, Buenos Aires (Argentina), September 2010

Ino, S. & Sato, M. (2011). Human-Centered Metal Hydride Actuator Systems for Rehabilitation and Assistive Technology. In: *Handbook of Research on Personal Autonomy Technologies and Disability Informatics*, Pereira, J. (Ed.), pp. 154-170, IGI Global, ISBN: 978-1-6056-6206-0, Hershey, New York

Jan, Y. K. & Brienza, D. M. (2006). Technology for Pressure Ulcer Prevention. *Topics in Spinal Cord Injury Rehabilitation,* Vol. 11, No. 4, pp. 30–41

Michlovitz, S. L.; Harris, B. A. & Watkins, M. P. (2004). Therapy interventions for improving joint range of motion: a systematic review. *Journal of Hand Therapy,* Vol. 17, No. 2, pp. 118-131

Sakintuna, B.; Lamari-Darkrimb, F. & Hirscherc, M. (2007). Metal hydride materials for solid hydrogen storage: A review. *International Journal of Hydrogen Energy,* Vol. 32, No. 9, pp. 1121-1140

Salter, R. B.; Hamilton, H. W.; Wedge, J. H.; Tile, M.; Torode, I. P.; O' Driscoll, S. W.; Murnaghan, J. J. & Saringer, J. H. (1984). Clinical application of basic research on continuous passive motion for disorders and injuries of synovial joints: A preliminary report of a feasibility study. *Journal of Orthopaedic Researche,* Vol. 1, No. 3, pp. 325-342

Sasaki, T.; Kawashima, T. & Aoyama, H.; Ifukube, T. & Ogawa, T. (1986). Development of an actuator by using metal hydride. *Journal of the Robotics Society of Japan,* Vol. 4, No. 2, pp. 119-122

Sato, M.; Hosono, M.; Yamashita, K.; Nakajima, S. & Ino, S. (2011). Solar or surplus heat-driven actuators using metal hydride alloys. *Sensors and Actuators: B. Chemical,* Vol. B-156, No. 1, pp. 108-113

Schlapbach, L. & Züttel, A. (2001). Hydrogen-storage materials for mobile applications. *Nature,* Vol. 414, pp. 353-358

Schrenk, W. J. & Alfrey Jr., T. (1968). Some physical properties of multilayered films. *Polymer Engineering and Science,* Vol. 9, No. 6, pp. 393-399

Tamura, T.; Kawarada, A.; Nambu, M.; Tsukada, A.; Sasaki, K. & Yamakoshi, K. (2007). E-Healthcare at an Experimental Welfare Techno House in Japan. *The Open Medical Informatics Journal,* Vol. 1, pp. 1-7

Tsuruga, T.; Ino, S.; Ifukube, T.; Sato, M.; Tanaka, T.; Izumi, T. & Muro, M. (2001). A basic study for a robotic transfer aid system based on human motion analysis. *Advanced Robotics,* Vol. 14, No. 7, pp. 579-595

Van Mal, H. H.; Buschow, K. H. J. & Miedema, A. R. (1974). Hydrogen absorption in LaNi5 and related compounds: experimental observations and their explanation. *Journal of Less-Common Metals,* Vol. 35, No. 1, pp. 65-76

Wakisaka, Y.; Muro, M.; Kabutomori, T.; Takeda, H.; Shimiz, S.; Ino, S. & T. Ifukube (1997). Application of hydrogen absorbing alloys to medical and rehabilitation equipment. *IEEE Transactions on Rehabilitation Engineering,* Vol. 5, No. 2, pp. 148-157

Wiswall, R. H. & Reilily, J. J. (1974). Hydrogen storage in metal hydrides. *Science,* Vol. 186, No. 4170, p. 1558